Praise for *The Lonely Patient*

"It is a revelation to experience serious and even terminal illness from the viewpoint of a doctor who is gifted in expressing human emotion in illness. It is like stepping outside oneself and viewing one's disease through the eyes of someone who understands and is a companion on the journey. Stein is a gifted writer and novelist whose prose is often poetry. . . . Recommend[ed] for anyone who has or will journey on a path as patient or companion. It is a healing book. The greatest gifts of this book are the insights and techniques it provides to the patient who is confronting the isolating emotions of illness."

—*Oklahoman*

"Michael Stein's language and powerful insight will be of great comfort when we or anyone we love falls sick."

—Annie Dillard

"*The Lonely Patient* is more than just a survival guide or owner's manual for those who are ill or whose bodies are broken. Recognizing that only clinical recovery can nullify loneliness, the author reminds us that a temporary escape can still be found in memory, imagination, and hope. Physicians and especially patients will find that *The Lonely Patient* makes very good company."

—*Journal of the American Medical Association*

"A superb doctor and sublime writer, Michael Stein captures the pain, the loneliness, and the terrifying arbitrariness of life with illness." —Peter D. Kramer

"[A] bighearted doctor." —Slate.com

"Stein has an amazingly observant eye, a tolerant ear, and an intensely human and permissive frame of mind. He has that rare quality of hearing the voices of both the patients and their tenacious illnesses. . . . [A] courageous text. . . . It is a commendable book that is aimed particularly, but not exclusively, at family members and others chosen by circumstance to nurture the sick." —*Providence Journal*

"Seeing illness from the patient's point of view sounds easy, but among doctors it is both difficult and rare. This excellent, empathetic book leaves me with one thought: 'Dr. Stein, why can't you be my doctor?'" —Anne Fadiman

"Stein has added a worthy topic to the canon of medical literature, and this book could serve as an inspiring guide to patients and their caregivers." —*Post and Courier*

© Sandor Bodo

About the Author

MICHAEL STEIN, M.D., is professor of medicine and community health at Brown University Medical School. He directs HIV clinics in Rhode Island and Santiago, Dominican Republic. He is the author of five novels, and *The Lonely Patient* is his first book of nonfiction. Stein lives in Providence, Rhode Island.

The Lonely Patient

ALSO BY MICHAEL STEIN

In the Age of Love

This Room Is Yours

The Lynching Tree

The White Life

Probabilities

The Lonely Patient

How We Experience Illness

MICHAEL STEIN, M.D.

NEW YORK • LONDON • TORONTO • SYDNEY

HARPER PERENNIAL

A hardcover edition of this book was published in 2007 by William Morrow, an imprint of HarperCollins Publishers.

HarperCollins books may be purchased for educational, business, or sales promotional use. For information please write: Special Markets Department, HarperCollins Publishers, 10 East 53rd Street, New York, NY 10022.

FIRST HARPER PERENNIAL EDITION PUBLISHED 2008.

Designed by Betty Lew

The Library of Congress has catalogued the hardcover edition as follows:

Stein, Michael D.
The lonely patient : how we experience illness / Michael Stein.—1st ed.
p. cm.
ISBN: 978-0-06-084795-1
ISBN-10: 0-06-084795-6
1. Physician and patient. 2. Communication in medicine. 3. Stein, Michael D. I. Title

R727.3.S756 2007
610.69'6—dc22

20066048118

ISBN: 978-0-06-084796-8 (pbk.)

08 09 10 11 12 ID/RRD 10 9 8 7 6 5 4 3 2 1

For Nancy,

Anna,

Oliver,

and,

of course,

Richard

Contents

Author's Note

My depiction of Richard here is true to the best of my recollection. In the patient stories, I have changed names and specific circumstances so as to protect each individual's anonymity and the privacy of their families.

The Lonely Patient

Prologue

My brother-in-law Richard was a sculptor. For almost a year, he had been having vague symptoms, bloody noses and spots of numbness on the roof of his mouth, which he blamed on mold in his studio, the vagaries of New England weather, and a lifetime of exposure to epoxy-laden materials. In late March, we shared our weekly midday meal at Johnnie's Luncheonette. We each ordered the usual. For Richard, it was onion rings, which left his lips shiny, and a chocolate shake. I had a turkey sandwich, mashed potatoes, and water with a slice of lemon. As we skidded home in the snow, Richard told me that he had a rare form of sinus cancer. He looked no different than he had the week before: wild-haired and thick-nosed at fifty, an expressive face with deep lines slanting from his nostrils to the corners of his mouth. His hands, gripping the steering wheel, were large and knotty. His size had intimidated me as a boy; now as then, his dramatic voice was overshadowed by his yet bigger ideas about almost everything.

I had always thought of Richard as invincible, so at first I didn't believe he'd understood correctly what he'd been told. Then I realized he had just told me the impossible truth. He seemed to have no more to say about it, and I had a sudden, panicky feeling that he was waiting for me to say something medical, to give him advice. But I was speechless, trying to collect what I knew about patients, cancer, the pervasive possibility of death, the experience of illness, and all the comforting things I'd told the strangers I had cared for in the five years since completing my medical training. I was a young doctor then and didn't know what to say. I didn't want him to die. I wanted to reassure my brother-in-law, offer him uncomplicated hope, as well as quell my already rising sorrow.

At the time, I was a general internist practicing at an inner-city academic hospital, treating patients with AIDS and heart disease, colds, asthma, and chronic pain. I saw some of these patients in their homes; some in my dingy, green-carpeted, yellow-aired clinic; and some on the overcrowded hospital wards. During these early years of practice, I had, perhaps, concentrated too much on how people become ill and how their problems can be "fixed," without trying to imagine what it's like to *be* ill. Whether out of self-pity or self-negligence, or because I was caring for too many patients in too little time, or because of the daily difficulty of seeing pain and terror, already I was no longer attempting to understand what my patients were communicating in gesture and expression, what was unsaid: why they were unhappy, what was bothering them most, how their bodies and lives had changed since they became sick.

Doctors feel obliged to strike the right emotional balance with patients: not too harsh, not too pitying. We are trained to hear and give grim news. From the first days of anatomy lab, we are inclined, even taught, to shut down. Rather than disclosing too much about ourselves—our judgments and contempt, our moods and fear—we are expected to keep our distance from patients. We deny that this distancing is a choice. Presented as a necessary rule, a requirement of composure, this self-protection disappoints patients interested in the interior world of their illness, the emotional life that's not on show. As doctors, we can't help being moved by the heartbreak of the individual medical moment, yet we never forget that patients are also examples of minutely studied biological processes. We refuse to be either impressed or surprised as we apply Joseph Brodsky's suggestion: "Keep eyes wide open, not so much in wonderment, or poised for revelation as in look-out for danger."

Early in my career, with the repetition of seeing twenty patients a day in a small exam room, I became like one of my ill patients: I had entered a world of black and white, a world without nuance, with only a two-word language—"sick" and "well." Gone were subtle gradations of color and dialect that an artist would appreciate. Then Richard got sick, and in the car that day I longed for a rich, fully formed vocabulary with which to speak to him about his illness and to help me understand it. I didn't have one.

The challenges of communication were well known to me. I had published two novels, been nominated for awards, and even taken up the subject of medicine in my second book. Richard bragged to everyone that it was the best book ever written about doctors, even though he had never read it. Still, before that drive from Johnnie's Luncheonette, I had never fully applied myself to the notion of illness, to traveling mindfully across its terrain. I was on automatic, doing my work with the cool precision of debridement. I spoke the language of medicine fluently but hadn't tried very hard to learn the idiom of the ill. I always thought that a doctor in good health could never quite appreciate illness. Each patient's emotions seemed just slightly out of my reach. I was inarticulate about the patient's experience of illness, but I was also holding back, in part because of my training and in part because I believed that I didn't have the right to ask or intrude.

My formal training in talking with patients took place at the end of the second year of medical school when I finally put my textbooks aside and began the requisite course on medical interviewing. Clinical information-gathering, I learned, has its own stiff anatomy and order—chief complaint, history of present illness, past medical history, review of symptoms. I memorized the list of topics to be covered when speaking with a new patient and watched my supervisor, a senior doctor, perform a series of interviews. Finally, the day arrived. My teacher escorted me to the bedside of an elderly woman with heart failure and sat in a chair behind me. Glancing down at my notes so as not to forget any important details, I spoke with this patient. Caught up in the excitement of my first encounter, trying hard to keep moving through my list of questions and at the same time to write down her responses, I sometimes neglected to talk slowly or to nod in sympathy, simple human behaviors that, I was to learn, patients depend on.

For years after this course, my interviewing style remained close to what I had practiced as a student. Only as my experience with patients increased did I begin to understand that trying to make a patient feel better sometimes only makes him feel worse. Intent on teaching me the basics in four weeks, my teacher and I never had the opportunity to discuss how difficult it is to use humor to disarm a patient, or how to foster an ill person's determination, or how to make a patient believe he still rules his own body.

Illness arrives, literally, out of nowhere. Newly ill, the patient immediately recalls the sick days of childhood, the afternoons asleep and the midnights awake, the disturbance of the natural daily rhythm. As an adult, illness makes him feel out of place, unaccountably absent, far outside existence. The patient expects illness to enforce a sense of restlessness and fluidity, but groggy and passive, he soon feels taken over, trapped, imprisoned. The patient expects to be dependent on others, but not the humiliation and indignity this sometimes brings. After a very short time, the patient wants to pretend he's well, to look the other way rather than face the new reality that the body can be hurt, attacked, scarred, that he and his body can fail.

Patients cannot put into words the extravagant difference between how they once imagined they would handle illness and how they actually do once it is upon them. Illness is repressive. The ill person forces himself to feel calm, to silence the high keening of distress. He is determined to keep his turmoil dormant. There is a palpable denial of emotion as the patient resists feeling anything but bewilderment. Yet emotions creep up and first emerge in the small clearings of idle time, and when they do, patients have difficulty identifying and describing these new and uncomfortable emotional states that have arrived with illness. The patient holds back, despite an urge for self-expression, because he doesn't want to complain and he doesn't want to worry himself; being completely candid makes him vulnerable.

That day in the car with Richard, the experience of being a doctor, of caring for the sick, and of trying to anticipate the ill person's sensations, thoughts, expectations, and exasperation became poignant and problematic and meaningful to me as it never had before. It's taken me the six years since Richard's death to grasp what it takes out of one to be a patient and how doctors and caregivers might help recognize and restore what is lost during illness.

That day, when I looked over at Richard, tiny tremblings in his facial muscles registered the aftershocks of telling me his diagnosis. I should have told him then that a patient's version of the world changes quickly when he becomes sick. He should expect the desire to continue an undisturbed, normal life to seize him harder each day. Whereas before, the patient may have lived a life dominated by the pursuit of happiness, suddenly his only pursuit is health. The ill person becomes a new "self." Yet in the midst of illness, the patient is unable to explain that his old identity has been transformed, stolen. He's been issued a new identity card, the diagnosis on it. He slips it into his wallet, but it isn't his, it isn't really him. I should have explained to Richard that the sudden tenuousness of the version of the world before illness would leave him with questions: When illness goes away, will anything resemble what I once knew? Will my life ever be the same after this? Will I ever feel like myself again?

Before Richard was sick, the idea that I could teach him anything was far from his mind, and mine. He was a natural teacher; he was the eccentric homebody who stayed in his studio and worked and listened to the radio, absorbing everything he heard and questioning most of it. *You* came to *him*. If I ever offered my opinion about art or finance or Boston sports teams—subjects he was expert at—I often felt like he was waiting for me to finish making a fool of myself so he could tell me what was what. He was willing to be surprised—he expected you to bring him things, pieces of information, gossip—but if you were going to discuss *his* subjects, you'd better come prepared. His wife, my oldest sister, and their daughter offered a not-so-gentle opposition to him, but I often didn't mind simply listening. Richard was exuberant, fierce, energized by talk and a challenge. Beyond embarrassment, he was difficult to shame for his bad behavior, his rudeness, or his increasing querulousness when the New England Patriots were losing and you were boring him. I was a little boy and Richard was twenty-nine years old when we first met, and although he was loving and generous to me, he also, later, teasingly doubted that I had ever really become a doctor: "How can an eight-year-old be a doctor?" he asked.

When he told me what the biopsies and scans had shown, I could tell he'd gotten a sniff of catastrophe and he hadn't liked it at all. He was quiet. As if he'd missed a turn, Richard took a long, winding route back from Johnnie's, driving down streets I'd never been on, past unfamiliar houses. Sickness, it occurred to me, is a foreign kingdom, an unrecognizable neighborhood. We might prefer to stay home, but sooner or later each of us is obliged, at least for a spell, to spend time in that other place. Perhaps Richard already sensed where he was going.

If illness is like going to a different, disturbed country, then the experience of illness—moving through that land—can be thought of as a kind of travel. It is an odd sort of journey because the sick person receives no invitation; he is suddenly, involuntarily, taken *there*. The sick person wonders with mounting anxiety: What am I supposed to do? What am I supposed to think? How long will I be forced to stay? Who can I talk to? Why am I here? Where do I go next? Illness puts him in contact with a primordial innocence that is dangerous during travel. There are many risks apparent. Some are obvious. Death is always lurking in the forest.

That the outcome of the travel is by nature ambiguous—one hopes for the best—in itself causes patients to live with vulnerability. Yet patients wonder what point there is in focusing on illness when they're just passing through. They're not settling here; they're just visiting, moving on, with their return tickets in hand. This change is dramatic. Once arrived, one is immediately ready to depart.

"One's native land has a sort of hopeless attraction, when one is away," D. H. Lawrence wrote, and anyone who is sick wants to go home again, to that version of the body he knew. Yet illness, like all travel, is an experience of jangling emotion, of moods and silence, of transitory feelings. In this new country, there is a different sense of time. Strict chronology is gone, replaced by what the physician and writer Oliver Sacks has called "personal moments, life-moments, crucial moments." Illness does not proceed by design; each step is unexpected and can come on suddenly.

In the car with Richard that day, the unbreakable silence made it a ruined, spoiled ride. Patients try not to linger on their anxieties and often can't or won't discuss them, yet at some point the undiscussed, unmentioned experiences of illness inevitably overtake them. Illness is far more than a diagnosis to be treated—or not. There is a particular alienation that illness brings. I have come to understand that the ill person's distance from others is the most profound experience of illness, and that this sense of other-ness—of loneliness—is more common in illness than any other emotion, and more dangerous and disturbing. And yet, perhaps, loneliness may also be alleviated. I wish I'd been able to give Richard the words and maps to move away from that place where there is no company, where nobody can follow.

I wish I could have told Richard the things I have said to patients who came later, and what I describe in the chapters of this book. I understand now that part of being a doctor is helping patients interpret new feelings and providing them with the faded directions back to their old lives or to the final and feared destination of death.

Understanding the emotional cascade of illness would not only help patients communicate with their health care providers, families, and friends but also provide them with the vocabulary necessary to describe their responses to illness. Patients may not be able to control the malfunctioning of their bodies, but they can have a chance to harness their ways of thinking. Illness that is articulated may lead to feelings of coherence and safety during stressful times and thus relieve the sense of loneliness. A patient's capacity to carry on is critically dependent on satisfying emotional needs for understanding, love, expression, and respect.

During this last decade, it has become commonplace to hear celebrities speak of their colonoscopies, for our leaders to have their hearts and prostates described in the newspaper, and for cable channels to broadcast close-ups of surgery. But even today the emotional side of illness is rarely mentioned. Thinking of Richard, I have realized that during illness certain human experiences are intensified, and four feelings in particular: betrayal, terror, loss, and loneliness. Patients must change their relationship with doctors and loved ones so that they are able to discuss these feelings and avoid being taken over by them. When we are ill, we are filled with a perplexed sense of difference—from what we were before, and from those around us—and too often this sense of difference is ignored at the risk of worsening isolation.

I want to tell Richard all I know now, offering him stories from my practice to get at what I've learned since his death. I want to tell him about my mistakes and the harm I might have done. Richard would never have wanted me to produce a standard series of clinical vignettes. He would want me to violate taboos, to trespass in my storytelling. He would not want me to focus on the history of a particular disease, or on a discussion of recent treatment advances, or even on the ethical problems that certain patient encounters bring up. Instead, he would want to hear about the more philosophical aspects of illness, about the meaning of illness, about illness as a different state, and about this idea of the "new self." We are interested in illness, either our own or others', because all of us will be ill at some time in our lives.

The Lonely Patient opens a new dialogue about our expectations of health and, after its shocking disappearance, of illness, our hopes, what we're capable of bearing, and our mortality. I'm convinced that Richard understood fairly quickly that he was about to be taken out of his daily routine and thrust into a new landscape. (He hated most landscape painters, except for the Hudson School's Ralph Blakelock, whose work was dark and swirling.) If there is even the slightest chance of understanding what goes on in the land of illness, Richard would want me to reach across the borders and bring back both a unique medical perspective and a writer's understanding of narrative—the story of sickness as experienced by the patient.

Part 1

BETRAYAL

As soon as the patient crosses into the land of sickness, she realizes she has been abducted. Many of her kidnappers are wearing masks and head coverings. They do not tell her where she's going. They carry unidentifiable instruments that could be weapons and move at a frantic pace in unrecognizable patterns. According to Oliver Sacks, she is in "No-land, Nowhere," and has "fallen off the map, the world of the knowable."

There had been no warning that Joanna was about to be stolen into a world of rioting, but when she began feeling pain, she understood quickly that she had been dragged into a fight she didn't start. She felt she had been taken over and knew intuitively that she had to protect herself from her illness. The feeling of betrayal intensified when she realized that there was no villain; rather, it was her body that had turned on her. The arrival of Joanna's pain and the illness that caused it interrupted the continuous state of her body in health. The foundation, the presumption of the music of the everyday, had been undermined. Her body was no longer an ally. When she was healthy, the availability of her body when she needed it had given her a sense of security and confidence. This assurance made it possible to play sports, make money, paint, climb ladders, make love. Her body had always provided a secure base, but once her body became sick, that security had been breached.

Joanna felt the world through her body: when she bathed, watered her garden, tipped back her chair, reached for a spoon, brushed her teeth, bowed from the waist, tumbled into bed. Like all of us, her unique, irreplaceable, and healthy body was the reference from which she made sense of every experience. Joanna took this so much for granted that illness felt like a betrayal. The world lost meaning, or at least she discovered that the meaning of her world was based on health to a greater extent than she had supposed. The meaning of life became simplified: I am my body, my body is sick, I am sickness. Life's stability had been threatened. The journey to the state of illness had been short and ruinous.

Often the first sign of the body's betrayal is the arrival of pain, and from the outset the most crucial fact about pain is its immediacy. Patients in pain think about nothing but that. They wake up thinking about pain. They go to sleep thinking about it. In *The Anatomy Lesson*, Philip Roth wrote that his protagonist, Nathan Zuckerman, goes "back and back to the obsession" of *not* being in pain. Patients in pain wait for it to go away a thousand times a day, and they wait for it to return. When it's gone, even for a minute, they can't remember it or re-create it. But when present, pain beseiges every thought. It "circles back on itself, diminishing all else," Roth wrote. Patients are enfeebled by pain, engrossed by it. In pain, there can be no sense of well-being. The idea of living forever never occurs to someone in pain, Roth wrote. Pain cornered Zuckerman, bound and gagged him, ground him down, took away any imaginative power he had. Pain is an unforgettable part of the state of illness, and any patient would agree to give up some of their life if they could escape from it.

Joanna came to my office one afternoon in the first week of December, a scarf wrapped high around her face, ice melting from her eyelashes where her breath had condensed. She hadn't been to my office in well over a year; at that last visit, according to the notes I kept in a purple chart, she'd had a nasty cold. Now she moved slowly, as if old, although she had just turned forty since I'd last seen her. She had an odd beauty: long arms and legs, dark eyes and eyebrows, dimples. She wore a silver ring on her left thumb and was dressed stylishly. Joanna's clothes had always been a tactile language of sashes and ribbons and tassels, fields of metallic and mesh fabrics separated by stitched borders, zippers, off-line buttons. Dressed for winter, Joanna wore black corduroy pants, a quilted scarlet jacket, an indigo crepe blouse, and black sneakers with dark red racing stripes and treads that climbed over the front of her toes. She stared at the new landscape painting I'd put up recently and at the wall-mounted blood pressure cuff and ophthalmic devices, her eyes stopping at the green examining table at the far end of the room. As she took off her coat and scarf, she observed the tall window, which let in northern light that changed as fast as the expressions of the patients I saw there every day.

"You have to help me" were her first words.

"I'll try," I answered.

As a young doctor, I found my patients' pain—the chronically swollen joints of rheumatoid arthritis, the muscle spasms of opiate withdrawal, the fluid that inflamed the lining of the lung—to be unknowable; it felt far away, and I was comfortable keeping it there. In some small way, I feared that if I recognized pain, if it came too close, it would somehow cling to me. Even if I was not afraid for myself, I did not like to admit that patients who came to me again and again for relief sometimes bored me. Pain is always new to patients but repetitive to doctors, Zuckerman knew. Everyone gets used to the patient's pain except the patient. It's always original to the patient.

After Joanna sat in the molded black chair between the side of my desk and the office door (all books and papers were pushed to the nonpatient end of my desk, clearing space for Kleenex and my phone), she told me, "Three months ago, I began to have this burning feeling across the bottoms of both feet." She was a lively and striking woman, but she sounded tired when she explained, almost apologetically, that she had been to see several doctors about her pain, starting with a podiatrist in early October. "He said, 'I want to X-ray your feet to make sure nothing's broken,'" Joanna reported. "'I want to see if there's a broken bone in there.' But I hadn't had an injury, and the problem was in *both* my feet. Of course nothing was broken."

"So what did the podiatrist tell you when he didn't see a fracture?" I asked.

"He said that as people age they lose padding on the soles of their feet. The layer of support thins. He thought thinning was the cause. He told me to cut out aerobics and to stop playing in my basketball league. He told me never to go barefoot again. He told me to go to the pharmacy and buy a big bottle of Advil, and if I wasn't better in a few weeks to come back." Obviously, pain had returned and stayed. Joanna shut her eyes and tried to make the pain disappear, something she'd been attempting for three months.

A body's betrayal tells you about a patient in its own strong way. I have seen the facial contortion of a ferocious headache and watched patients gasp and yelp and clutch themselves and collapse and vomit from crescendos of pain. I have witnessed patients carrying pain into the exam room as though it were an additional family member, melancholy and dangerous. I have watched patients pretend their pains had subsided, protecting me from what they were sure I couldn't handle. Some patients hide their pain to provide me, silently, with an understanding of what their lives might be about.

What has impressed me over the years is how much patients *hate* their pain. There is real outrage. Pain makes the future look dismal, and it's impossible for the patient to remember when it was bright. The body's betrayal is mystifying to patients, no matter how many times I go over plausible explanations. Whatever enlightenment I provide, the only *meaning* of pain for the patient is abject and unambiguous failure: one's body has not remained well. Regardless of the cause of their pain, the natural state of the betrayed patient is bitter disappointment.

For the endless line of strangers and longtime patients who come to my office, I am the source of usable information. Often I feel that I am of little or no use to patients, although they think otherwise. At least I can offer this: together, we loathe illness. Together, we can make a show of solidarity and defiance and concede little to illness and pain. I have a tendency to take the measure of a patient quickly and to decide silently whether I am likely to be of help in this way.

"So you took a break from sports," I said to Joanna.

My office was brightly lit by the winter glare, and the forced steam made the pipes rattle. As usual, the room was too hot, and the carpet, glued to the floor, lifted in the corners. The sun came in silver and made things look different, harder, than in the summer.

"Not completely. Not at first. I gave up aerobics completely; the teacher arrived at least five minutes late to every class, and that drove me crazy anyway. I didn't think my pain was from basketball, even with thin padding. I couldn't stop *basketball*. It was only once a week, Sunday afternoons, but it's been every week for ten years, right?" Joanna's athleticism, her toned body, was a point of pride to her.

"I started playing fewer minutes. I tried to jump less. I used ice after every game. I still wanted to be in the league, so I took lots of Advil. I thought I'd live with the pain. But it's worse."

Joanna didn't care about my advanced degrees; the only qualification that counted was whether I could make the pain go away. How did she attempt to make me understand her pain? She didn't at first. It would have been like defending her life's work as a landscape architect; it was simply who she was by the time she came to see me. She didn't immediately have the words. She didn't want to be indulgent, but she wasn't indomitable either.

I asked her to tell me more about the pain. Sometimes it was a hot line of agony that focused on the underside of the first joint of her big toes and the balls of her feet, but it never migrated up the leg or involved her hands. Other times it was a ticker tape of pinpricks, an electrical storm under her skin, amplified and staticky. Joanna was bitter and bewildered. She had no other symptoms; with tight brown curls, a tiny mole over her left eyebrow, and perfect teeth, she was robust and impressively muscled, with strong hands and no underlying medical problems. She had two children, and her work involved topiaries, bushes she groomed into animals and figurines; her height must have helped her handle shears. There was no reason to expect her to have foot pain; she had no history of foot problems as a girl, although ten years earlier, in her late twenties, she had been a rock climber. Despite the pain, she had driven to my office in a Honda with a manual transmission that required her to use both feet.

"You have to promise to stop basketball." I said these words respectfully, solemnly, with the inevitable sense of loss built in, but also including an expression of optimism. "It might help." I could tell that she was worried by my undertone of disapproval.

After Joanna stopped playing basketball and the pain didn't go away, she never again seriously considered that it was her fault, that she might have prompted the pain in some way. Pain came on after a day of exercise, or after a day of avoiding exercise. It started an hour after she lay down or an hour after she started moving in the morning. It was worse after dinner and during either high or low barometric pressure, and it could be so severe that it would wake her from sleep—and keep her awake. She ascribed her troubles to shoes, to sandals, to going barefoot. She blamed inflamed nerves, crooked bones. What made one day worse than the next? Too many stairs? Not enough vitamins? Driving the car or sitting in a chair too long? Swimming? She had always been a good student, and for about a year she figured if she understood her pain, she could overcome it. Joanna looked in every direction for explanations. She borrowed anatomy books from a friend. She searched the Internet for "burning feet" and "pain in feet." Then she gave up on science. Who cared about science? She had studied her feet so many hours that the wrinkles on her soles reflected the wrinkles on her face.

She tried to cure herself. Pressure on the instep helped, for a while. Keeping her feet raised and keeping her feet active offered some relief, sometimes. She put foam pads in her shoes for support, changed heels, and tightened laces. She kneaded the undersides, the topsides. Massaging for hours a day helped, but it was not practical. She was caught up in anguish. She used ice packs of every kind, blocks in blue plastic, loose cubes in wet washcloths, and crushable activated chemical cold. She tried bed rest. She wished for winter, to walk shoeless in snow all day; maybe that would help.

After ten minutes in the supermarket, she described each step as "walking on knots" and "having socks made of rope." The pain had come on gradually, inconspicuous from day to day, until it had become an established, relentless reality. More recently, it had taken over her daylight hours, and if her pain started in the morning, despair in wild proportions arrived by noon. Her voice trailed off. Her face was flat and weary.

"Three months of pain that comes on nearly every day is a lot to go through. You must have quite a high tolerance for discomfort," I said.

She was unconvinced by my praise. "When your children have pain, you wish it were yours. But now that I have my own pain, I'm horrified." She was distraught. She wasn't sure what to say without apologizing for her inarticulateness.

In the two weeks before her appointment with me, pain had kept Joanna away from a good part of her previous existence. She spent much of her day trying not to think about pain, but her mind was stuck on its arrival, its departure, and its return. No sooner would she think, *Oh, I'm not in pain,* than it would reappear. She began to think that *thinking* about pain brought it on, that if she could stop thinking about it, it would be over. Pain became her job. Like Zuckerman, once in pain, no other predicament was even conceivable. Not fighting with her husband, not having difficult clients, not taking care of her elderly mother, not chasing a ball down a beach, not guiding a baby down a slide. I had always thought of Joanna as a woman who could think of many things at once—the beauty of the snow as well as the fickleness of the Dow Jones—and despite her pain, I imagined it was possible—in fact protective—to make use of this ability. But other than thinking of her pain, or trying not to think of it, Joanna felt mindless. After three hours of pain, she could not converse civilly. She was no longer alert to the world around her.

In *The Anatomy Lesson*, Zuckerman has an undiagnosed syndrome imprinting a candelabrum of agony across his neck and shoulders. "Had he kept a pain diary, the only entry would have been one word: myself." Pain keeps a patient in the present because the part of him that is most alive, the most vibrant and immediate, is his pain. Patients in pain care about nothing but the pain they're in; they think only of themselves: What will make this worse? Will there be more? Is there something I should do, or stop doing? When a patient speaks of being "in pain," it is not unlike a felon speaking of being "in prison." And like many who are sentenced, the more a patient tries to work his way out, the further he works himself in. All that is left is pain, confusion, and a feeling that the body has betrayed him. The patient has trusted his body up to that point, taken good care of it. Why has it turned on him? This sense of betrayal is too enormous to manage—it swamps the patient, causing him to feel useless.

Pain often comes as a surprise. Because it is usually limited, because it is almost always finite, patients are not overwhelmed at first but instead are angry. With the slow discovery that it can become chronic, that one can slide, without transition, into what the novelist and memoirist Harold Brodkey called "a gray, oily, uncertainly lit sleeve of almost complete pain," the sense that the body is betraying one increases. There is the experience of pain and the knowing that it will continue. Surprise turns to disbelief, then anger and rage, and finally depression and despair. There is something appalling about waking every morning knowing that pain will soon return, that it is close and coming fast, that one will be alone with it for another day, unable to stop it. There is a descent toward resignation and helplessness.

Joanna had tried to map her pain in order to find her way back to painlessness. As Brodkey wrote, "Pain does not have to be charted unless one is determined to escape: and then it is charted only so that one can find a way to its edge in order to see the world again." Brodkey understood that pain was a different state: "It is not metaphorical or a figure of speech or a conceit to say that as that knowledge [of my illness] grew to occupy the center and the periphery of my attention, whatever else I knew seemed unimportant, and was, in a geographical sense, forgotten."

I start each workday expecting sadness. Sometimes the best I can do for a patient is to ask, "What is it like?" This question is a way to slow down a patient's rushing thoughts, to let him describe what might be accomplished during his visit. This question allows the patient time to gather himself and try to explain what he's feeling. Yet the sociologist Elaine Scarry has written that pain drives out language, that a person in pain cannot find adequate words. Pain is, literally, unspeakable. Pain's expression is a matter of arrhythmic, non-uniform syllables of moans and screams. Its dimensions are radically private, prelingual, unattached. Each person's pain is like nothing we have ever heard before. Pain destroys language; sounds of distress are its only form.

Scarry suggests that the only way we can describe pain is from memory, when things have calmed down. I have heard patients describe pain as tearing the body into pieces or lighting it up, as pins and needles, as a paper-thin incision, as depthless scorching, or as the sensation of being bitten all over, like rats gnawing at your neck, a sliver of bamboo slid under a nail, a bayonet prodding, Joanna's coarse rope socks, a sharp-beaked bird hopping all over your limbs and joints. Pain is God's arrow. It is like an echo in a house without furniture or curtains. Pain is a "despotic and possessive hostess," wrote the seventeenth-century philosopher Alphonse Daudet. It tells you where to sit and whom to talk to. But when pain has been rendered powerless, our descriptions must be untruthful. The metaphors never capture the true experience. Patients in pain are not prepared to measure their pain accurately.

Attempting to share our pain stories is a little like trying to describe love, an equally interior state. Love hurts, we say. This pairing is no accident. Both love and pain are invisible, yet people in pain can date each moment of pain just as they can recall each moment of love. Pain is a record of felt-experience, but it is necessarily inaccurate, dependent on previously heard formulations couched in a narrow, already existing vocabulary that doesn't capture its truth. Screams convey only a limited dimension of the experience. The poets have tried for centuries on the love front: "love is like . . ." they write again and again. What is pain like? Virginia Woolf wondered: do the sprain and the headache resist language?

Just as there is no general theory of love, and every episode of lovemaking is unique, love at least has an object. I am in love *with him*. Pain is different; it is a physical experience of negation, a sensory "against," according to Scarry. There is no commonly understood meaning in the language of pain. Pain is the birth of language, or many languages, each patient speaking his own. Of course, the language of pain, like the language of love, is untrustworthy and always shifting. "To have great pain is to have certainty, to hear that another person has pain is to have doubt," Scarry has famously written. I would say: to be in love is to have certainty; to hear that another person is in love is to have doubt. Think of the private sounds of lovemaking—cries and groans. Are they expressions, relatives of pain? What would their translation be to another language? No such thing exists.

Joanna climbed onto my examining table, did a heel-to-toe to push off her sneakers, then pulled up her knees to take off her short black socks. She sat there, slow to move, not unhappy at the chance to rest. She let her feet dangle so I could have a look. Her feet were scrubbed, the nails clipped short, no polish. To the trained eye and to Joanna herself, her feet looked symmetrical, decently proportioned, normal. They were pale and narrow and cool, the soles a reptilian white-pink. Odorless, thankfully. (Some patients have feet that smell like wet leaves, berries, bicycle seats.) She looked down at her feet and watched me handle them, as if thinking about what they were made of.

"Thankfully it's not every day, but when the pain is bad," she said as my fingers rode over the heel, the arch, the ball, as I asked her to curl her toes, "I think I'm going to pass out." She swayed a little bit as I felt around. When I tested her reflexes, she gave little, helpless, normal kicks.

Invisible, unexplainable pain is a matter of enduring wordlessness. In some ways, pain induces, perhaps demands, wordlessness. The patient believes that any attempted description is necessarily unhelpful, or an acclamation of weakness and petulance. She senses that her pain is always doubted, partially disbelieved: if the pain is invisible, it must be trivial or imaginary or evidence of hysteria. Speaking of pain makes patients feel sullen and embarrassed. Yet, as a doctor, I depend on language and trust the patient as a reliable narrator. I need to believe that patients are capable of accurately expressing the most resistant aspect of material reality. Sometimes I need to coax them into clarity and their own interpretations. That is my job, to create an etiquette of permission, made more difficult by the fact that each patient speaks her own language of guilt and self-blame.

The problem, of course, is that pain is often invisible, not caused by injury or obvious trauma. It is subterranean, interior and remote. It is alarming and yet, in a certain way, unreal to anyone but the patient. To a doctor, pain cannot be made to disappear, because it never appeared in the first place. There is no way to confirm a patient's complaint, there is no scan or blood test or examination finding, and this is frustrating for me and other professionals. We believe in cancer because we can see it. Pain can't be confirmed or denied, it must be believed. Indirectly, it can be heard through shrieks and moans or visualized through grimaces. Without these clues, doctors are unaware of its existence in the patient in front of them. Many times in my career I have been in the presence of a person in pain and not known it. Yet, to the patient, nothing is clearer, louder, or more apparent.

When we think of pain, even chronic pain, we think of injury. Descartes had a bell-ringer model for pain. The harder one pulled the ringer, the louder the bell rang; it was a linear relationship. Analogously, the worse the organic disease, the worse the pain. This model is easy to accept, but it is wrong. Horrendous war traumas are sometimes painless, while minor injuries can be awful. An inflamed hangnail can wake you at night simply by brushing the wrong way along a pillowcase.

When there is no escape from pain, we externalize it—I am being stabbed, choked, burned. Pain is not created by us, but is done to us. For Joanna, it was as if a hammer were coming down on her foot. As if a truck were driving over her toes. As if her bones were broken and the jagged ends were sticking through her skin. For her, pain was the slowed and continuous state of being acted upon. How could Joanna expect her body to get better if she was under attack from a force she couldn't even see? Pain was simply her body's response to the invisible damage, and because it was invisible, it was terrifying. In one of Scarry's examples, if a nail went through one's foot, it would have shape, length, and color; unlike the pain it would cause, the nail would exist, we would accept its reality. But Joanna had only meager words to describe her feelings; she could not show me any penetrating nail. When you are ill, when your body betrays you, you say, "Something is hurting me," when actually the something is you.

"Does anyone else in your family, your parents especially, have similar pains in their feet or hands?" I asked. Taking a family medical history is standard, but usually patients have already beaten you to this area of inquiry. Illness makes us think of our parents, the insurgency of their genes, their illnesses, their brain and body chemicals that explain our own. Yet we rarely know our parents' personal histories, let alone their medical histories, in any close way until a problem arises, and only then do we collect details to learn if we've been cursed in similar ways. Symptoms sometimes point to one's genetic origins, and when medical problems emerge, patients naturally look backward for what's been passed on, especially those patients who are ready to blame their parents for anything.

"I have my father's sense of humor and his reluctance to laugh at anyone else's jokes," Joanna began. "I stroke my left cheek with my thumb the way my mother does when she's thinking. I have her voice, and I worry like she does. But as far as I know, neither of them, nor anyone else I've asked, has had the nerve misfirings I have."

Her parents, alive and well, offered nothing that would provide a shortcut through the list of possible causes. Heredity had no part in explaining Joanna's pain.

Joanna's ability to deal with pain was impressive. She never babied herself. She put a high value on her considerable indifference to aches and fatigue. But as hard as she tried to imagine feeling better, she couldn't get anywhere. She had lost her belief that good health is natural.

What is one supposed to do when *this* happens? That is the question of all illness, and after months of pain, Joanna had had enough. She had tried to be positive: *This will go away.* She had imagined improvement with good American naïveté and optimism. She had thought: *Now it's time for recovery.* But when the pain returned, the old questions returned: *How can this be? What am I going to do?*

I didn't know what was wrong with Joanna's feet. But to

lessen her fear I said, "We see this," a comment that hid the fact that I had never seen anything exactly like her case. Doctors always try hard to sound natural, as if nothing surprises us. I fled to another room and sat and thought about her eczema, her recent tetanus vaccination, and her aunt's history of colon cancer and unsuccessfully tried to put them into an explanation for her foot pain. When I returned, I ordered blood work to make sure Joanna didn't have diabetes or an anemia that would signal some systemic illness. She was frightened by the possibilities.

I had little to offer her immediately, either diagnostically or therapeutically, as I waited for the blood tests to return, and I told her so. Together we decided she might visit a physiatrist I recommended, a specialist in rehabilitation medicine who might come up with some form of immediate relief. I suggested a higher dose of ibuprofen, to be taken with food. I understood that Joanna wanted to "drive pain out ... the way a clapper knocks sound from a bell," quoting Roth. She wanted to hear the sweet ring of her discharged affliction.

Joanna came back to see me the third Tuesday after New Year's, about seven weeks after her previous visit. From my fifth-floor window, I could tell that outside it was windy, with flags stirring and loose paper flying past. The latest snow had melted. It was morning, and the light in the room was thin. Under her bulky black parka, Joanna wore a plain white blouse and gray slacks. The alphabet colors and shapes that I knew her for, the spontaneity and play and awareness of life she had always displayed through her clothing, were missing. I had spoken to her a few days after her last visit to let her know that her blood results were normal, but I had not heard from her since. In her time away, she had become, like Zuckerman, an apprentice to pain; she had subjugated herself to its will. She reported that the situation was close to intolerable.

"It's so much worse."

"I figured as much, or you wouldn't be here."

There had been moments of reprieve during the past month, but Joanna considered them false signs. She was thankful whenever she got an hour of relief, but it only allowed her to recalibrate. Sensations pushed in and out and in all directions. Wild-eyed and tentative in my exam room, she opened her fists to let some of them go. Joanna was full of sensations. Her body had betrayed her; she didn't know how or why. Would she ever be able to trust her body again? The point was for me not to betray her as well.

Unusually, she had nothing to report about her children, her husband, or her trouble parking in one of the far-off hospital lots. She was desperate to talk about the pain. The physiatrist I sent her to had squeezed an area near the balls of her feet, and when she cried out in pain, he diagnosed her with a Morton's neuroma, Joanna told me, speaking quickly, frantically. The neuroma was a small, nonmalignant, fibrous growth trapped between the small bones of her foot. She'd read about neuromas and chronic foot pain; if she educated herself, everything would be solved. "It can be removed with surgery," Joanna informed me, "but the physiatrist had suggested cortisone shots instead, to shrink the neuroma, so I could avoid surgery." She had agreed to three injections, two weeks apart, which themselves had been painful.

"The first shot gave me pretty good relief, right up to the second shot. The second gave me a break too, but I wanted it to last longer. Being pain-free made life seem simple for a few weeks. It reminded me of the old days." The reprieve faded even faster after the third injection, and her old pain returned, but with a change in character and intensity. The original small, burning area on the soles of her feet now involved the tops of both feet and all her toes. The physiatrist admitted he was baffled by the persistence of her symptoms.

From the beginning, Joanna described her pain as having a thermal dimension, a burning. Throbbing, it had what pain researchers call a temporality. (I've heard other patients describe pain as pulsing or beating or flickering.) Joanna now also experienced a spatial/restrictive aspect. Pain pinched and pressed and gripped her feet. Joanna, an enthusiast of popular culture, described her feet as being crushed in a George Foreman grill. She told me that she was willing to have her feet cut off to gain a brief respite.

Without asking, without moving to the examination table, Joanna took off a new pair of white running shoes (which, I'd noted, she left untied), showing me the plastic orthotic—a specific heel and arch support—the physiatrist had molded for her. She removed her socks and began to massage her feet. It was as if she had muscle cramps she couldn't relax, couldn't stretch out. The leather of the table, buckling with her slow, deep motions, made a sad, halfhearted sound.

"I've swallowed a thousand anti-inflammatories, rubbed in the prescribed creams. For the last month, I've seen your physiatrist and an acupuncturist a friend recommended so many times my calendar is blank except for visits to them." Nothing had made a difference, and the pain had progressed.

I realized that at the last visit, when I had seen the pain in Joanna's eyes, I had slowly backed away. This time I reached out, took her hand and held it, letting the pain run out of her for a moment. I had failed her. I had not done the thing demanded—freed her from pain—or sent her to someone who could. I sat on a low, rolling stool at the level of her knees. Dangling, resting on my palms, Joanna's feet were nearly weightless. I checked the cortisone injection sites near the small toes on either foot: there was nothing swollen, nothing red signifying infection. I was quiet, making room for her unease and questions and helplessness. As she bent over in her chair and started to replace her socks she stopped, her forehead touching her knees. She was shaking. I stood and put my hands on her back to comfort her, and I felt her sputtering breaths, the vibration of her firm shoulders. The rawness of her pain was obvious, and stunning.

For weeks Joanna had sensed her own weakness, had tasted

hesitation. She had gotten further and further away from life as she knew it—its speed and celebration and comfort. She limited her time driving; she thought twice before shopping at the mall, where she might have to walk a long way between stores. Before every errand, she considered what she needed to do to get through it: double her medication dose, pick the right shoes. With every action she thought, *When I get home, I will pay for this.*

In pain, Joanna drove fast when she drove at all, fighting down feelings of wildness. She screamed at other drivers; she screamed at her kids. I wondered if Joanna didn't at times become hysterical at home. I knew I would have. I wondered if she was careful around her children—trying not to yell too much—but if when they left for school she threw herself to the ground and howled. I have heard patients' cries so piercing that I couldn't tolerate the sound and had to move away.

She didn't want to scream; she wasn't like that. Plus screaming conveyed only a limited dimension of the experience, she told me. Her anger was directed at the two lumps she called feet. The pain was so bad at times that she forgot what she was doing, where she was. Joanna, who was the kind who took on blame, confessed to me that she now rarely called her sister (also a patient of mine), rarely saw her, and felt badly about it. Her body was now the enemy. Pain had established an internalized sense of adversity for Joanna: my body hurts, my body hurts me. Her feet had become the agents of her own annihilation. Joanna's pain was self-announcing. It was an atrocity that her own muscle and bone had inflicted on her. I wanted her to have mercy on herself. I admired her ability to be strong and stubborn.

She was not merely suffering pain—it had broken her. Pain was an incomprehensible occurrence; it defined helplessness. It made her passive because it was so punishing. Pain was an extreme condition. There was no ceiling on pain. It rose; it seemed never-ending and brought along nausea, fatigue, and dizziness. During illness, very little matters, Brodkey noted, but what does matter is of unassailable importance.

Patients sometimes try to take the nineteenth-century physician's view of pain: the greater it is, the greater must be our confidence in the power and energy of life to overcome it. But if they are not convinced, the sense of betrayal grows deeper. Pain went a long way toward explaining Joanna's moods and erratic behavior, she said, her "forgetting" to call her sister on her birthday.

Joanna was angry; she wanted to take charge, to overcome the power that pain had over her. Her first impulse was to make no adaptation to pain at all. I'd thought that Joanna was crying when I touched her back, but actually she was quivering with rage. She had lost the modulation of a mature, confident, reasonable, authoritative adult. She told me she used to be someone with manners, someone who was decent, courteous, decorous, upright, and civil. Now she begged. Her voice was out of control when she screamed at me, "I'll do anything."

I knew she had been trying to outwit the pain. She pretended it wasn't hers; it belonged to someone else. She'd lost any kinship with her body. Joanna wasn't Zuckerman. "When he felt pain he'd pretend instead that it was pleasure. Every time the fire flares up, just say to yourself, 'Ah, that's good—makes me glad to be alive.' Think of it not as gratuitous punishment but as gratuitous reward. Think of it as chronic rapture, irksome only inasmuch as one can have too much even of a good thing. Think of it as the ticket to a second life." Joanna wanted to whoop and laugh and pretend her life was free and full of possibility. But she couldn't because it wasn't.

A part of Joanna suspected she might be crazy. Visits to four doctors had brought no explanation for her pain. Although she sometimes feared that whatever was wrong would kill her, she was pretty well convinced that her pain wouldn't; it was not a terminal illness, just pain. But some piece of her wanted it to be a dreaded disease so she could explain herself to people. If she looked ill—lost weight, lost muscle tone—people would believe her, accept her illness. To appear healthy almost seemed a liability when dealing with people who saw her wince. That her pain made no sense, that it was incoherent and hidden, made her crazier still.

"How's this for a delusion," Joanna said. "When I was up last night at one in the morning, sitting on the side of the tub with my feet in an ice bath, I started to think that it's possible that the pain inside me will get used up. If I put up with this suffering for long enough, there will be no more left in my body. The well will be empty. Even as I was thinking this, I knew it wasn't true. It couldn't possibly be true. But I thought I'd ask you anyway. Why not? What's the harm in asking? If I let my pain stay really bad, won't it go away faster? Then I'll never need to know what caused it because it will be gone. But if my pain won't go away, what am I going to do?"

The strain on Joanna was tremendous. Daily, she made a bargain with herself: she would suffer just one more day if tomorrow might be different. I wanted to tell her that she was correct, that her pain would evaporate, as if that would settle it. Most people go through life dreading that they'll have a traumatic experience; Joanna's had arrived without the noise, the crash, the tragedy. Her illness was a test.

Away from their medical visits, patients think about illness secretly and guiltily. When the body has gone wrong, it's as if they've *done* something wrong. I've come to understand that my profession grants me the specific moral authority to explain that illness is not of a patient's own making. I need to explain to patients that they are not fundamentally unlovable; they are always ready to believe the worst.

Rather than offering Joanna a new answer I didn't believe in, I occupied myself with things I knew: checking her blood pressure, feeling her pulse, listening to her lungs as she slipped on her thin white socks, her sensible, loose sneakers. Trying to distract myself, I noticed that the floor hadn't been buffed or even mopped. The green-specked linoleum was gritty. I wondered how long it would take the cleaning crew to come through after I put in a request. I imagined the very first day Joanna realized her pain was not going to go away, the day she awoke betrayed, not expecting to encounter illness. She expected a world with the same dangers, rewards, and disappointments as the day before. But then it all changed. Her illness was unbidden, willful, unpredictable, and for nearly half a year she had struggled to maintain her life against it.

"Morton's neuroma isn't right," she announced. "I know it isn't." She had a cracked nail, and she fingered it, looking up at the ceiling, trying not to cry.

"I agree."

"I've had four doctors look at the same set of facts. Each decides which facts he likes. Meanwhile, I'm starting to imagine a life when I can't get around, when everything is brought to me—by mail, by truck, hand-delivered by my husband and daughters."

The ill take it for granted that everything about illness is knowable. They are powerless not to accept this belief, or why see a doctor at all? A doctor merely needs to hear a patient out, perform a physical examination perfected by years of training, do the necessary tests, and get on to judgment—a three-act drama with a compelling finale. But often what we tell patients is agonizingly imprecise. With Joanna's pain, I was struggling with a problem I hadn't come across before.

Joanna admitted she'd also been to see an orthopedic surgeon in addition to the physiatrist and the acupuncturist. Each doctor she had seen focused on a different possible cause, each looking within his specialty and ordering tests particular to that specialty. One believed that Joanna's problem originated far from her feet, in her spine; another suspected the culprit was the elevation of her left hip. As the list of possibilities grew, so did her pain ("the top of my foot is asleep, and the bottom feels as if I've been out dancing all night in shoes that were too tight"), and she took the lengthening list from one doctor to the next, changing her story slightly each time, trying different descriptions with different doctors. She then brought the opinions back to me to sort through with her. Her pain was frightening, oppressive. What is this pain a symptom of, what's the *real* bad news, Joanna wanted to know. Every pause in response to her questions felt long and fatal.

"I want to put all of these professionals in one room and listen to them argue with one another," she said. "Could you arrange that? Even if the answer at the end of the argument is, 'Your pain will never go away,' it'll be a relief." Only by putting all of her doctors together in one room would we see how many she had consulted—evidence of her desperation, her seriousness. This way she would be sure that the answer had not slipped between the cracks of the specialties. Joanna was convinced that if any one doctor had listened with perfect pitch, had tuned into her frequency instead of his own, he would have gotten to the bottom of her pain. It was impossible to accept that there wasn't an answer. Because if the doctors she'd been to didn't know the answer, who did? When I could give her no immediate answer, she began to think of her pain as a punishment.

Pain has always had religious overtones. "Life is the place of pain," says the *Bhagavad Gita*. Suffering is the central metaphor in Judeo-Christian thought: the test of faith in the story of Job, the sacrificial redemption of the crucifixion. C. S. Lewis, in his book *The Problem of Pain*, suggested that without pain we would not take note of God's power, nor have any reason to glorify him. The early religions imagined Hell as a place of unavoidable and incalculable physical pain. The modern Hell involves psychic punishment, disembodied torture without end. Both Dante and Milton, in emphasizing that in the "doleful shades" of Hell pain is permanent, differentiate punitive pain from the moral cleansing associated with transitory discomfort. Pain worsens only because one refuses to know God or bliss. Suffering brings refinement; pain is the punishing route to a transformed self. With the introduction of ether and the sedative gases, religious writers called such anesthetic practices a violation of God's law; they believed that God inflicts pain to strengthen faith and teach self-sacrifice. But for us moderns, perhaps less concerned with the purifications of Hell and more concerned with the present, bodily pain trumps psychic pain. As Oscar Wilde wrote, "God spare me physical pain and I'll take care of the moral pain myself."

Joanna wanted to give her pain *significance*. What did it mean? What was she hiding? What was she revealing? What was she betraying? Nobody could make her believe that she'd had pain because she deserved it. She was not the kind to accept that she'd had too much good fortune, that because life had been going so well she was due for hardship. Pain had not arrived because she needed it or because she could take it. What made her so resentful was that she couldn't anymore. A year before, Joanna had no premonition that something bad was about to happen.

She did not wish to be a suffering person for any theological or psychoanalytical reason. She did not accept that pain is the body's sinister twin, taking its revenge for whatever crimes the mind has supposedly committed. Trying to determine a cosmic philosophical answer to *Why me*? was merely an attempt to find a pattern in chaos. The best Joanna could come up with was a hardened "If we're born to die, I suppose we're also likely to get sick along the way."

Joanna had finally reached Zuckerman's opinion of his pain:

> It's not interesting and it has no meaning—it's just plain stupid pain, it's the *opposite* of interesting, and nothing, *nothing* made it worthy unless you were mad to begin with. Nothing made it worth the doctors' offices and the hospitals and the drug stores and the clinics and the contradictory diagnoses. Nothing made it worth the depression and the humiliation and the helplessness, being robbed of work and walks and exercise and every last shred of independence.

From the very beginning, Joanna had played thought games. Was it better for her pain to be constant, like slipping into quicksand, or for it to be erratic, taking her by surprise? Which would more quickly break her spirit? Which would crack the frame of her human dignity more completely? In what manner would pain establish itself as master over her faster? Which way would allow her to feel more hopeful? Joanna had never contemplated suicide. She was convinced that constant pain was easier, even if its coming and going dashed all hope.

"I have given a name to my pain and call it 'dog,'" announced Nietzsche. "It is just as faithful, just as obtrusive and shameless, just as entertaining, just as clever as any other dog." With all of Joanna's tests normal, I continued to tell her, "It's not . . . it's not . . . it's not this, it's not that," but I couldn't say, "It is . . . I'm sure it's this." For months I had no useful information for her, no diagnosis. I spoke softly in my ignorance, but all my words sounded too loud. I wanted to say something absolutely correct that would be protective of her, that would guard her. I felt I was betraying her. Undiagnosed pain makes both patient and doctor feel impotent.

"I don't know how I'm going to continue like this for the next thirty years," she said.

During my early years of medical training, I distrusted pain. When I saw patients in the emergency room with a crushed fingernail or a swollen knee joint, I tended to base the legitimacy of their pain on what I imagined I would feel if similarly afflicted. I thought of personal injury law as practiced in the courtroom, where the very extent of pain is in dispute and the clear expression of it engenders monetary reward. And I was in the business of inflicting pain when it was necessary: I drew blood, passed catheters, took biopsies. But this pain was transitory and obvious, and the particular features and persistence of Joanna's pain left me weary.

Over the days following Joanna's latest visit, I did my own research about foot pain, speaking to several neurologists, reading neurology texts. I called her and asked if she would take another test, an unpleasant one. "I'm ready," she answered. I sent Joanna for electromyography, an EMG, a test in which needles inserted into the nerves of her legs sent electric pulses toward her feet to determine if the conduction of electricity was slowed by disease and if the sensations in her legs were normal. The results were inconclusive and, by their lack of diagnostic specificity, suggested a condition I had read and heard about but had never seen: tarsal tunnel syndrome. Her tibial nerves, originating in the lower spine and passing along the length of her legs, were getting mechanically trapped in their fibrous sheaths as they passed along her ankles, and this pinching of the nerve fibers was causing her symptoms, I believed. The tarsal tunnel diagnosis wasn't certain in my mind—certain features of Joanna's complaints fit (the type and location of the pain), while others didn't (both feet being affected at the same time and the improvement with activity)—

but it was clear to me that her pain was *nerve* pain, neuropathy. She didn't seem to have a systemic illness that was appearing first in her feet; I believed the cause was nearby, at ankle level.

She returned to my office a week after the EMG. When we shook hands, her palms were slick. The longer it takes for a doctor to give illness a name, the more a patient is afraid of what he might say; the fatal illness is always *last* on the list, the hardest to arrive at. Of course, there are also patients, particularly those who are asymptomatic, who don't even want to know the name of their illness. If they don't know the name, it may not happen, it may not exist. They want to remain ignorant. They don't probe, even after tests. All they want is another long, calm day. They believe that the news will sink them fast. And if they don't ask, they won't be told. Tarsal tunnel seethed with unspecified threat for Joanna until I explained it several times. Sometimes what I say is casually forgotten by patients, but Joanna heard my diagnosis clearly, significantly. Lucy Grealy, in her memoir *Autobiography of a Face*, recalled that when she received her cancer diagnosis, "it was as if the earth were without form until those words were uttered, until those sounds took decisions, themes, motifs. There may have been thousands, millions of words uttered before those incisive words, but they had no meaning, no leftover telltale shapes to show they had existed."

Joanna had an illness that was *felt*. She knew that some illnesses are silent, some perhaps imagined, and that to name illness is to unmask it, to have it out in the open. She was ready. The tarsal tunnel diagnosis actually relieved her. I had given her pain a name. The name transferred her condition from a speculation to the concrete medical plane, that is, to an experience with a worldly reality. Her disease was not life-threatening, but it was unseen; she had been navigating the shadowy world between illness and health. Naming it meant that her pain could be known to others, that there could be no argument. She was willing to accept any diagnosis as long as it meant she wasn't dying, and when I told her she wasn't, she ticked off plausible milestones of improvement: a full night of sleep, a morning when she didn't have to ask her husband to rub some life back into her toes. Still, I was surprised when she said, "It would be easier if I were disabled. At least I wouldn't have to pretend for other people, try so hard to act normal, or need to put on a cheerful face." I hoped that by handing over the words "tarsal tunnel" I was giving Joanna a break.

In giving her a diagnosis, I gave her permission to talk about things that perhaps no one had wanted to hear about before: the mysteries of her pain and her modes of self-composure. "It's like tending to a fire," Joanna said. "I'm engaged by it, but also too busy. I have a purpose, but it's also work. It clears the air of other smells. It coordinates the rhythms of my brain." Pain enforced a sense of patience in Joanna. But at the same time, I knew what Joanna was thinking: *Will he find my pain a dreary and self-pitying recounting? Will he abandon me?* She knew that doctors have to be interested and patient and curious to see the puzzle of pain through to the end.

She moved around the room, breathing deeply. Even at this moment, her pain was so intense that she couldn't be still. I thought: *Maybe she didn't scream out loud because she was yelling inwardly.* Joanna alone knew her pain, but it was unheard, unsounded. She tested the words I'd given her in her own voice. If no one heard a patient, did it make her feel even more alone? She tried not to fall as pain bloomed across her feet. She tried to control herself, tried not to be disgraceful.

Pain requires something of the patient, and it requires something else from the doctor. For me, embedded in my reaction to seeing someone in pain is wishing that person immediate relief, although pain also sometimes raises my curiosity—could *I* bear it? How bad *is* it? The medical task, written in its very oath, is diminishing or, better, eliminating pain. None of us wants to watch a person in pain. Because I had learned the pharmacology of pain control, my compassion could be translated into a series of actions and prescriptions. As she paced I explained to Joanna what I knew about tarsal tunnel, but more importantly, I told her that something decisive had happened to her and that her pain might not be counteracted by regular doses of philosophical thinking before any of the new pain medications kicked in.

Joanna was angry because she knew that pain is not ennobling in the long run and because tarsal tunnel is not easily fixable and would therefore be around for a long time. She had every right to be angry. Her body was not the way it was supposed to be; she was not going to be the way she thought she'd be. She was going to have to be someone else. She had been betrayed.

When I believed that she was not angry with *me*, I asked, "What's the worst thing about all this for you? What makes you angriest?"

"The doctors who were quick to dangle theories with a cure attached. Doctors who said, 'Here's what you need to do,' and when I did what they said and it didn't work, they walked away," she answered.

Joanna had been given five or six different explanations and unsuccessful treatments before she came back to me, and my goal was, in its simplest form, not to leave her hanging. Patients need to believe that suffering is treatable. Otherwise, although our compassion may be strong, it is likely to yield a quality of despair. "If you don't fit into a diagnostic box, you start to think maybe you're making it all up. That crazy feeling I told you about," Joanna said. Pain is always a psychological state, whether or not we appreciate a proximate physical cause. With Joanna, I had alleviated the suffering that came from "not knowing."

Abandonment of the hope of future recovery was what most frightened Joanna. It was early February, and in my office she asked, "What do we do now?" I wanted very much to give her good news. But she wasn't looking for good news. Surgery for tarsal tunnel is simple, but its results are unpredictable; many patients show no improvement in pain, and some get worse, I told her. She was willing to have any kind of surgery if she would experience *a fraction* of the pain. She wasn't smiling when she said this.

Like most doctors, I suggested narcotics instead, stronger analgesic medications, in an attempt to buy time, to let time heal her. Joanna had been using the pills I prescribed at the last visit, but they hadn't helped much; at best, they'd cut the edge briefly, even as we'd moved to stronger brands and dosages—Vicodin, codeine, Percocet. I knew nothing more about pain relief than Joanna it seemed. I couldn't answer Joanna as to why the pills hadn't helped much, and it left me—and her—feeling unsatisfied. With pain, the orthodox may not work—it may alleviate nothing. Since you can't afford to reject anything, every possibility must be explored. She didn't want a groggy, heavily drugged life, however, and she described her mind on morphine as fogged glass. She wanted to work outside of a trance. For a hundred reasons no one could explain, the medications I had given her had failed. She'd had experience with medications up to here, and it hadn't paid off.

At times I believe my power is greater than it is; I imagine a patient's pain has been vanquished sooner than it actually has. When their pain has not been eliminated, I understand some patients' urge for alcohol, for stupor and sleep. Consciousness seems worthless if pain is around. It wipes out irony, invention, history. Drugs cover and disguise pain, although patients know, given the limitations of medication, that it's coming back.

So, knowing that my prescriptions had failed, I let Joanna talk. Doctors need to urge patients to speak. We do this by providing encouragement, making comments like, "I admire your strength," or, "I don't want you to be a hero." Patients don't know that they need to speak, and indeed they find it difficult to speak; pain has made them mindless. I've spent entire mornings with people in pain who have never said a word. It's too much for them. The mind etherizes itself against whatever is too enormous to bear.

I have seen patients in so much pain that they have forgotten their own names. But they never forget those who help them. Doctors hold the world in place for patients. We establish the legitimacy of their claims. We offer human contact and concern apart from the private context of sympathy. The best we can do as doctors is turn ourselves into a reflection of our patients' pain. The doctor's job is to make pain shareable. The impossibility of sharing or articulating the feeling of pain is a component of the pain and a contributing factor in its essential horror.

When a patient's symptoms do not improve rapidly, she senses that others, certainly her doctor and often her family, are disappointed in her. And rightfully so. The cruel truth is that her illness represents work for others. A patient can sense impatience in those around her if she complains. Complaining is the language of pain. When complaining, a patient doesn't know whether to make herself sound worse or better than she is. When complaining, she's actually crying out, "I'm sorry," because she is contrite about her pain, about being physically unfit, about asking for help. The impatience of caregivers contributes to her shame, while her conscientious silence in the face of real reason to complain leads to a feeling of exclusion. In this way, the sick person begins to exist at a slight remove from common life, and she travels further and further away as time goes on. Illness injures her sense of importance, and each patient is lonely in her body's betrayal.

In talking with Joanna, I discovered that she did best when she stayed busy. Immersion in mindless activity made her forget her pain, her self-focus, any self-pity. Like all good patients, Joanna didn't want to cooperate with illness. I wanted to applaud her frantic scrambling for solutions. We tried another medication, gabapentin—designed specifically for certain neuropathies, it might improve her symptoms, I thought, when combined with simple aspirin—and low-dose opiates. She played with the timing of her pills: one of these with breakfast and lunch, two of those just before bed. Everything had to go just right to keep Joanna moving, and when she had a good day, she examined the good fortune of her method the way a painter adjusts the borders of brightness and then steps back and appreciates her work.

Joanna also joined a chronic pain support group, which she called the "Weeping Is Not Worth It Society." She reported that their meetings were full of hilarious sympathy, exhaustion, and a raw appetite for pain-free life. She wanted to keep up-to-date on her options: behavioral therapies, biofeedback, local hypnotists, novel implantable devices. It was stunning to Joanna that some of her comrades had given up the fight. These few, she reported, were frightened of change and familiar with and reconciled to a certain level of long-term misery. They had assumed—among friends and relatives—a kind of heroic stature, and they enjoyed being praised for their courage. The writer Reynolds Price, describing his three years of neck and back pain following spinal surgery, might well have been describing Joanna's group when he wrote, "The hurting has so nearly *become* us—become the whole core of our present self—that the thought of finally dismissing it from us feels scarily like desertion." Yet underneath the seeming inability to change, the seeming masochism, was another worry, I suggested: her group-mates did not want to try treatments that might make their pain even worse.

During those hours when Joanna was not in pain, she remained in fear of its return. This prescience is what I call suffering. I think of suffering as pure energy, a raw force that has to be released or tamped down—like opening up a pipe that contains a high-pressure gas with an angry desire to escape. Pain becomes unbearable when the patient believes it will get worse. The anticipation is deflating. Perhaps as Joanna's doctor, I could help with this. If suffering is subjective, perhaps it can be modified by subjective actions—touch, talk, prolonged bonding. Through intimacy and caring, doctors may enter the patient's imagination. Pain renders the patient intensely susceptible to the unsolicited arrival of loneliness. I once read that the function of literature is to serve as an antidote to suffering through depiction of our common fate. The function of doctoring is to serve as an antidote through the expression of the patient's suffering, through putting language to the dry roar of pain.

At our visit in March, I tried to get Joanna to tell me about pleasure, the moments far from physical pain, and she did: running silk through her fingers, feeling the skin of her beloved, or the breeze through a screen window. She was willing to tell me what she'd learned from pain. She told me that pain had taught her what well-being is, what cowardice is, what it's like to be sentenced to hard labor. As her doctor, I could not cross the distance to Joanna's pain. But she excused me. She told me that the distance from chronic pain to painlessness is astronomical.

She advised me of her latest techniques. For Joanna, balancing the pain in her feet with the normal sensations in the rest of her body was a negotiation. She tried to isolate the feeling in one foot or the other and then list the parts of her feet that *didn't* hurt. This constant physical awareness was too much at times, but she used her secret philosophy: she told herself that she was lucky to be alive. At times she tried to forget everything she knew, all biological laws, and attended to the world around her rather than her own desperate moments. When her pain was almost comic in its intensity, her humor returned. Incessant pain had caused her altruism to reappear as well: since nothing could be worse for her, she was freed up to help make life better for others. She started to volunteer at a center for people with dementia.

Each day, the patient in pain lives the child's dream: someone will come in the night and rescue me. When told that they may *not* get better, but that they won't be abandoned, patients are sometimes willing to accept pain at a certain level, as Joanna was. Although my second pain medication, for instance, helped only with her pain at night, she could sleep again. Pain is an alarm, but doctors can change the pitch.

For those in pain, life gets smaller and smaller and smaller. They confront pain, worry and wonder about it, fight it, treat it, ignore it, hate it, try to figure it out. They chart the fluctuations even when they can't make sense of it. Patients talk to their pain, reason with it, and when it doesn't respond, they're met with indifference. A sense of alienation only adds to the pain. Doctors can lessen the urgency of a patient's pain by giving it a place in the world. We counteract the natural force of being swallowed by pain. If we do not acknowledge pain, as Scarry notes, if we deny its presence, we double pain's annihilating power.

The betrayal of the body comes in many forms. Richard didn't have pain like Joanna's; instead, he had a different form of discomfort, a teasing, untouchable numbness on the roof of his mouth at the edge of where his tongue could reach. The diagnosis didn't make him angry, as I had expected it would. He almost seemed to accept the news as one of life's predictable sadnesses we all pass through. As a veteran of the art world, it was his modus vivendi to absorb misfortune and move on.

His calm about having cancer was a surprise. Richard was frequently angry at the world. I'd seen him explode at airline ticket counters. He would walk away feeling he had been right for the screaming scene he had made and the managers he had called in. He often felt betrayed or unappreciated. Curators had left him out of shows; reviewers had overlooked or belittled exhibitions he headlined. He berated the world, then he started on me. I had betrayed him plenty of times in small ways—canceling, at the last minute, a plan to meet him at the Museum of Fine Arts or forgetting an article I had promised to bring him when I visited—and he would magnify these personal slights into evidence of either my disrespect for him or my personal weakness. Only his cats could be counted on. Cecil and Alice lounged on his desk when he worked, snuggled against his thighs when he watched television. He cuddled with them constantly—they lay on his chest at night—rubbing his knuckles over their skulls as they butted him. When he fed them half a can of food each, it was a ceremony of some importance.

He had not experienced cancer as a betrayal. He viewed his cancer, lodged deep in his skull between his eyes and extending into two sinuses, as fate. There was no malice or motive involved with getting cancer. No one had done him wrong. He was physiologically incapable of blaming himself, as Joanna had. It was just bad luck.

In the car driving home from Johnnie's Luncheonette, the air had stopped in my throat after hearing Richard's news. I wondered if he remembered the conversation we had had not long before, when he'd told me that at a certain point after turning fifty years old, if you listen to your friends, most of what they talk about is their body's deteriorations, the wear and tear. People are obsessed with feeling well, he'd said with disdain.

I always preferred to ride as a passenger with Richard, because anyone who takes the passenger seat is admitting that there are things they can't control, and Richard had always seemed uncontrollable to me. I waited for him to ask me a medical question about his cancer. Every question he'd asked me for twenty years—about a movie we'd seen, about the strength of the Celtics' first-round draft pick—smacked of contentiousness, and he didn't ask questions at all if he thought he already had a better answer. He made even his natural curiosity seem hostile with his challenging tone. There were no safe subjects with Richard, and none where his opinion could be avoided, where his words didn't penetrate and disquiet.

"Do you know anyone with cancer?" he finally asked facetiously. It was his way of putting me down a little, of saying that it was time for me to just admit that I wasn't *really* a doctor. But I knew from my sister that he was always impressed that I put on a tie to go to work and that I spent all day with sick people. Knowing that I was not an oncologist and had little knowledge of cancer treatment, he was baiting me to show off my professional stuff. But I understood that his question was also a way to release me from feeling that I had to help him: how could I be expected to help him if I didn't know anything about cancers? Under all of these meanings, he was asking me to find out all I could about his rare malignancy. The incompleteness of what I knew about his form of cancer, after all my years of training and practice, was not so much embarrassing as sad for me; I wanted to do well for him.

With most patients, I discuss intimate matters, but I am not a friend. I am responsible for their symptom management or cure, but we are not close. There are always self-imposed limitations to our conversation. They know almost nothing about me, and don't ask. I am laconic. I make people ask before I offer predictions and probabilities. They know what they see: I am thin, moderate, calm. I am usually very good at handling the combination of grave danger and constant neediness that is the awful presence of illness. I'm fluent in the language of medicine, and I'm happy when my patients get better. It took Richard's illness to make me feel powerless.

Perhaps men don't know how to comfort each other except in a black-humor, shared-trench kind of way. After his teasing question, Richard's expression was serious. I didn't quite understand back then that it's hard to know what's troubling a person unless you ask; all I knew was that asking felt like an intrusion. Richard wasn't someone who prayed, but he must have been struggling and scared, and he figured that I knew what people go through, or at least knew what was coming, and might be able to tell him. Richard drove on with his diagnosis while I simply looked outside at the world and its shadowless brightness speeding by.

Betrayal often involves the disrupting of expectations, and when we are healthy, we expect to remain that way. Imagine breaking an ankle: even after the cast comes off and we again put weight on that ankle, we don't "trust" it. We have been betrayed, and we don't expect much; the ankle may do what's necessary, but no more. We perceive the ankle differently from before; we have a new image of it as weak. To relearn trust, we need to rehearse it. We take long walks before we attempt skipping rope. We have to adapt to this new relationship with the betraying, untrustworthy body. Illness is a disturbance of trust.

Health is familiar, predictable, reliable, and, we hope, enduring. It provides a sense of orientation. Illness is a break in the established, continuous sameness and comfort of health. Betrayal arrives without arrangement, unpredictably, spontaneously, and carrying danger. It is a threat, and we are vulnerable. It has revealed a secret about us. Personal worth and value are undermined. All of us idealize our own bodies (even if not every piece of them), so we are deeply disappointed by illness. We are strong and vigorous one moment, helpless the next; we have power one moment and are without it the next. We take account of our assets and resources, but when betrayed, we feel useless. Our emotional equanimity depends on health, its safety and stability.

With the discovery of cancer, patients immediately face choices: radiation, surgery, chemotherapy, surgery plus chemotherapy, chemotherapy preceding or following the operation, radiation beamed or implanted. Sometimes there are so many choices that they are more frightening than empowering. Patients are asked to make decisions at a time of great stress, under great pressure (*Do something now!* the patient thinks to himself), and before they've had time to gather information and consider the implications. If they delay—and often patients have more time to make a choice than first appears to be the case—they appear indecisive. For weeks I pestered Richard about a surgical date, but he told me to leave him alone. Then he went ahead and scheduled surgery without telling me.

The five hours of surgery were an ordeal. Richard's cancer was deep behind the bridge of his nose. In that part of the skull, thin, brittle bones establish a series of delicate chambers, sinuses. It took the surgeon hours to peel back the skin of his face, to approach the bony perforations where nerves snuck through, looking for the cancer that coated the nerve strings hiding in the dark. The surgeon looked down and in and scraped soft tissue from the naked skull's microscopic corners and cul-de-sacs. (Afterward, Richard's sinuses would become dry, echoing caves, sensitive to and inflamed by cool inhalations.)

During his recovery from the sinus surgery, Richard's greatest fear was pain that couldn't be controlled. Some people look forward to pain and its daily demands, and others take secret pleasure in distress and are married to their bodily aches. Richard wasn't scared of surgery, only of the pain to follow. His first line to every doctor he met was, "Don't hurt me." He said it as if he were invoking a curse. Then when a new doctor arrived, he would turn to me, his witness, and say, "Some doctors like to hurt people. That's just what they're like. That's not what my little brother-in-law here is like, though, and he's a doctor." I'd offer a weak smile in the midst of Richard's weakness. I didn't mind being his medical credential. It seemed the least I could do.

In the hospital, he wasn't shy about playing the cancer card with the nursing staff when he had postoperative pain. Unlike Joanna, he had no interest in bravery, or in even pretending that he was brave. He asked for relief without hesitation. He didn't fear retaliation from the nurses for pushing them to bring his pills on time and for asking the young doctors to increase his dosages. He had never feared punishment for his desires; at home he had few impulses that he forbade himself.

Richard had always loved the pleasant buzz of pain pills. I knew this after the first root canal he ever had, when he called me to say his dentist was a cheapskate and hadn't given him enough codeine tabs and could I call in a few more for him. I consented, but I wasn't amused. He was familiar with his own bad character. He never needed much of an excuse to partake in narcotics, but his sinus surgery was a good one.

One of Richard's natural facial expressions was that of false misery. Even when things were going fine, the corners of his mouth turned down as if everything in the world was just too much for him. He was pained by bad ideas, disappointed by banality. When he was well, it was hard to decipher what might be bothering him by just looking at his face, and in the days after surgery, his sinus pain did not smoothly relent. He claimed it was incrementally more painful when it returned after a break of a few hours.

He welcomed the dazed doziness the pain pills delivered. He was not sentimental: illness was about rising to the challenge, but only to a point; then one gave in to its eventual erosions and forgot as much as one could along the way. He didn't believe that illness offers a larger way of being alive, or gaining a greater knowledge of life's dangers. He wanted to get out of the hospital as soon as he was able to walk the ward. He didn't like the looks of the patients, but he also didn't like to see the visitors, soon to be spouseless or motherless; their loved ones on the surgical oncology ward clung tenaciously to life, but every day someone lost their grip. He felt the breath of melancholy. He was spooked.

When he returned home, my sister had converted the living room into his sick room because it had the largest television. The living room was reduced to its constituent grays and whites and browns, stone and biscuit, the tones of the Chinese art he loved and collected so passionately. The bathroom next door became a medical room, a health center. It contained bottles of pills and tubes of cream, cartons of gauze pads, rolls of adhesive tape, disinfectants. It smelled of talcum powder and urine.

Richard lay on the gray leather couch in front of the built-in bookcase, facing the television and the impressive multi-paned window that looked out over the sloping front lawn. He was eye level with the line of evergreens at the end of the lawn, and past them, he could see the lake across the road.

Richard made use of his stockpile of pain medications. He kept a small skyline of brown bottles on the table in front of the couch and on his desk in the studio, where Cecil and Alice wandered between them, batting them around with slow paws, before settling next to the warm radio with its buttons and dials caked with clay and wax from Richard's work. Richard had collected a prescription from every doctor he'd visited in the weeks leading up to surgery, and from the dentists too. He had an addict's attitude toward opiates: they brought shameless joy. He had smoked cigarettes for thirty years and quit a few years before his diagnosis. But he still sucked on Nicorettes, a long-term replacement for his cravings, and cellophane carcasses littered his studio and living room and bedroom. Weeks after surgery, he sometimes sucked on the narcotic Percocet instead of a Nicorette, and who wanted to stop him? The expression on his face seemed to say, *I'm a little high, and I know that I'm naughty, and you would be too if you were in my position.*

Before, Richard's visitors always came to *him*. When he was ill, this didn't change. He never asked anyone to stay away until he was back on his feet or able to visit *them*. Illness did not change his character, only his tempo. He didn't suddenly become mystical or spiritual. He didn't believe that pain offers some masochistic pleasure or morbid secret gratification. His recovery was filled with menial, mundane thoughts. Clothes or pajamas? What sort of lunch? Shoes or not? Simple decisions to reacquaint himself with the old life.

We think of illness as a smooth narrative unscrolling with a clear beginning and finish. One gets to a destination, an answer, a view, a final clarity. There is a driving force, but the travel of illness is actually circuitous, looping, and full of descents and ascents. It is not a straight line into this strange country. Illness proceeds in knots of time, tangles of fear, clumps of misery. News arrives in the form of a doctor, and for Richard it came during an appointment the week after surgery, when all the pathology results had been reviewed.

Richard liked his surgeon, a tall, thin man with a slow manner of speech that reminded Richard of his own Louisiana roots. He liked this surgeon in good part because the surgeon liked him, got a kick out of him—Richard's willingness to say or do the unexpected regardless of taboo—and was never dismissive or unfeeling. He never said, "There's nothing to be worried about." He jousted with Richard, teased him back, babied him a little. Richard trusted this doctor to be straight with him. That day his surgeon reported that they had not been able to remove every bit of his cancer. Some tumor was still in there, having spread along tiny nerves and into the tiny caves of his cranium. Radiation might arrest the progression, his doctor suggested.

That evening Richard and I watched TV in silence. There is

a shyness in sitting with the ill, a kind of shared sorrow. For patients to speak, there is always a resistance to be overcome first, a resistance built on embarrassment and exhaustion. I realized that I was studying Richard, not only because I was concerned with how he was doing but to see how he handled his news, how he would muddle through. I wanted to see if psychic pain would turn him away from the world. I was not unhappy with the quiet in the living room. As a doctor, I was used to constructing nervous shields that protected me against patients' upsets.

Early on, I had often assumed that my patients' emotional states would last much longer than they actually did. Richard had experiences other than illness competing for his attention and emotions. He had a new art show coming up; he had work to do. He took the position that it was perfectly acceptable to proscribe and bury the experience of illness. Not all bad news had to be processed. That evening I began to think about the natural resilience of all patients. Sick, they believe that nothing else will ever be quite as intense, nothing else will ever really matter after surviving and cheating death. Yet even when illness leaves a permanent paranoia, patients move on because what choice do they have? They must each possess an innate mechanism to dampen the sting of adversity. They move on to other things in life that *do* matter, and they appreciate the sweet joys that come along. This hadn't occurred to me earlier, perhaps because I had not lived long enough to be ill myself.

Illness rarely allows a connection with simple happiness.

The moments of happiness during illness often consist of feelings of relief, of escape from apprehension and from whatever is wrong. Joy is measured in negative terms—no pain, no weakness, no bad news. But Richard refused to be unhappy. Maybe he took the long view: he knew the shock of diagnosis and even surgery would fade eventually if he allowed it to.

As guests arrived, Richard's policy was never to admit to terror. His light tone was deceptive. When he first told his friends the news of his cancer, he made it sound almost silly, as though he wanted to create a sense of disbelief for others and presumably for himself. If he really had an illness, it was easy, slight, not terribly serious, easy to dismiss or distance into something less than formidable. He valued a certain mastery of himself. Flannery O'Connor wrote in a self-deprecating letter to a new acquaintance, "You didn't know I had a DREAD DISEASE didja? Well I got one.... I am bearing this with my usual magnificent fortitude," which was approximately what Richard said when I overheard him on the phone, telling his art dealer about his diagnosis.

Richard didn't feel that his body had betrayed him until later, when his cancer returned eight years after his operation, this time not in his sinuses but in his brain, and he began to stumble. By then, I had begun to think of his cancer as a slug destroying the tomato plants in my garden: it was impossible to imagine how something so tiny could take down something so huge.

Part 2

TERROR

For many patients, there is the fear of the unknown, the fear of making wrong turns and bad choices, of being separated from companions and of traveling alone, of humiliation, of getting into something they can't escape. The sick need to make sense of things, yet they have a constant sense of incomprehension and a feeling of ignorance that some days is exciting, hopeful even, and other days is anger-provoking and frustrating. During illness, as when traveling, one alternates between adrenaline and submission. Every day one lives the biology of anticipation: *What's next for me?* Still, during illness there is really only one choice—to proceed or not to proceed. And not proceeding is giving up.

Within moments of shaking his hand, I noticed the bump on Luke's forehead, just above his right eyebrow. Painless unless tapped smartly, it was first spotted by Luke's wife, Sheryl, who'd noted it over a month before but didn't mention it until a few weeks later. She admitted to being too uncertain about its origin (had he banged his forty-year-old forehead doing woodwork in the basement?) to say anything. He had no other symptoms and had never had any bony growth like this before; he probably wouldn't have done anything if Sheryl had not finally said, having overcome her worries about overreacting, go see your doctor.

The day Luke came in was unseasonably warm for June in New England, topping ninety degrees. The air conditioner had gone out in my Volkswagen, and even at 8:00 A.M., when I was headed to work, my shirt stuck to the car seat. The sky was white. In my office, there had been a brownout the previous night. I opened the windows to let in a breeze that was mostly imaginary.

I had recently watched a PBS show on plate tectonics and the long-ago formation of the Himalayas. It described how subcontinental plates pushed together until one buckled, bringing about the underground cataclysmic upward bursting of a new ridge. Under the smooth expanse of Luke's forehead, his bump looked like K2 must have when it first formed, elevating the hard crust, carrying its ground cover (skin) past the surface. Luke's brow had become mountainous, or at least hilly. I suspected it must have been a bone cyst, a skull-based osteoma, but then again I had never seen anything quite like it on that part of the head—nor had I heard about something growing as quickly as Luke's wife described.

"If it's been getting bigger, I think you'll need to see a surgeon about that," I suggested. He nodded, but I could tell he was only half-listening to me and was instead hearing the faint sound of singing, an overplayed radio tune, from speakers in the hall outside my room. I looked out the window and saw a jogger running on the street below and wondered if Luke saw him too and wished he were that man, escaping.

In the minutes that followed, Luke asked me none of the obvious follow-up questions, the whys and wherefores, the full list of diagnostic possibilities and prognoses that give patients their minimal sense of control. Some patients set out to learn how their bodies work—or do not work—so they can understand what goes on inside them and can better fathom what has gone wrong. It takes a certain amount of bravery—they know they'll be frightened by worst-case scenarios at first—but later they may be able to subdue their terror with sure knowledge.

"Surgeon?" Luke asked, as if waking up. He seemed awkward and reluctant. I gave him the surgeon's name quickly, as if neither Luke nor I wanted to admit that danger was coming; this wasn't anything to dwell on. "Whatever you say," he agreed. I saw Luke shrug, all his desperation in that gesture.

He was a large, muscular man who dressed in jeans and black T-shirts in the summer and jeans and oversized turtleneck sweaters in the winter. He always wore green Vans sneakers, like the skateboard kids I saw hanging around the 7-Eleven. I thought of Luke as lighthearted and carefree; he played street hockey with his kids and was close to his father, who, I'd heard more than once, had retired from the insurance business. Luke was a graying frat boy whose two favorite four-letter words were "free beer." When I thought of him, I thought of a big dog eager on his leash. Luke's wife, Sheryl, on the other hand, was small, crisp, and energetic, with short hair and tiny earrings and fingernails. She had a notebook and recorded anything I said. When I mentioned the surgeon, she was cold-eyed, neither sentimental nor ironic—she simply wrote it down. She wasn't near tears; she knew her job was to remain calm. I expected her to speak up as dark thoughts entered her husband's mind.

"What kind of surgeon?" she asked.

"Probably a neurosurgeon, since the growth is on his forehead." My words sounded too loud. I said "his" as if Luke weren't sitting right next to her. I could see that she was ready to ask the questions he couldn't, but she held back, waiting.

"Don't look at me like that," Luke said to his wife.

Luke must have felt that he had been dragged to see me by his overprotective wife for a tiny swelling that he would have been happy to pretend didn't exist, but hearing the word "surgeon," he realized he was getting into something deep and difficult and he might need to retreat. He shrugged again, which meant he was confused and in need of help but still trying to maintain the appearance of nonchalance. In a ninety-degree medical office, when the sweat is collecting on a patient's eyelids, a shrug is an acknowledgment of discomfort. It was as if Luke had been given an impossible equation to solve, but fear had already started to deprive him of clear thinking. All he *could* do was shrug.

With diagnosis pending, a patient knows he is about to learn something; he is on the verge of what the memoirist Paul West called (when he was about to have a pacemaker implanted) becoming "a pupil of your own mishap." But Luke was not ready to learn anything yet. He was too scared for that, and he had not yet settled with the idea of illness, despite the bump above his eyebrow. It was too fast. His eyes flickered with disappointment. But what did he expect? Of course he knew that tests would be needed (although he felt healthy) and that other doctors were likely to be called in; that was why he had put off coming to see me. Once he'd made the appointment, he knew there was the potential for hearing difficult news.

The possibility of illness brings on thoughts of vulnerability, worried-over endings, the worst outcomes. For most well and intact people, it's more comforting to think of a sudden and definitive event—a heart attack, a stroke—when they think of illness. A dish dropped on black granite. Slowly advancing, their trajectory unclear, chronic illnesses are too abstract, their borders too fuzzy, to see past them. Luke's bump had been around a while (chronic), but now that I had raised the possibility of surgery, his predicament seemed acute. He'd asked me nothing, but given that his forehead was swelling outward, I knew he must have been thinking about where it started: brain tumor.

Luke's shrugs suggested more than desperation. He was slipping out of control, though so gradually that it was hard to even notice. Terror is subterranean, a tremor, a distant, imagined happening. But it is also close-up and as definable as surgery. Even before a patient knows exactly what he's dealing with, he senses how ill equipped he is. Everything he fears and dreads is coming, and all the phantoms of the mind are at work even before a diagnosis is offered. "Don't look at me like that," he'd said to Sheryl; Luke felt watched. He was being tracked by some enormous eye. He had crossed a line of fear that he never even knew was there, and suddenly he was observing *himself* too, closer than ever before.

Does terror concentrate the mind? No. It distorts everything patients hear, or think they've heard. Listening to me, patients sit absolutely still, bodies rigid, as they learned to do in school years before. But Luke's mind was not still. Terror is a fast-forward button. It speeds up all thoughts. At the same time, terror slowed Luke's ability to make sense of anything. He was quiet with me because it was easier to not ask questions than it would have been to stop once he started. Terror has no reverse gear; you can't take it back. I could tell that Luke barely recognized his own voice when he spoke.

Patients watch doctors as carefully as they can without looking like they're doing it. They hunt for signs, trying to decipher what the doctor has not yet told them. A slight tilt of the head, the rise of an eyebrow, a slow deep breath. They want to be comforted, to be drawn back from the sense of expulsion that begins with the first sign of bodily betrayal and becomes magnified by a first medical appointment.

Sheryl took me aside at the end of our meeting. "Luke wants to make a good impression with you. He cares what you think of him." Patients want doctors to admire them. They worry about being thought of as cowards. As they try to be brave, they are sometimes overcome with fear. They employ what Harold Brodkey called "bravery tactics," which include politeness, discipline, and acting strong-willed. I saw this in Luke too.

"I think your husband is probably nervous about his forehead. No one likes to go to the doctor," I said. "He was just very quiet. But I can understand that. I worry, though, that I'm not being clear when a patient is so quiet." Sheryl told me that Luke was afraid to ask too much, that he feared making me nervous or overcautious if he asked too many questions; he didn't want to make me feel that I needed to prove myself.

I understood that Luke felt about his visit with me the way I dealt with my son's schoolteacher: although I was skeptical of her, I needed to be gentle or, I feared, she might take it out on my child.

Luke called me two weeks later, after his visit with the surgeon, to discuss what he thought he'd heard, to keep me in the loop, and to get my opinion. The surgeon, baffled by the bump even after several radiological scans, wanted to do surgery, and soon. She'd thrown out a lot of words I had to help Luke pronounce—plasmacytoma, eosinophilic granuloma, myeloma—all of the nightmarish names that stood in for the "C" word, cancer. "There's no way to think of this that seems right," he said. I could hear a haunted quality in his voice.

"The surgeon didn't convince you that you need surgery?" I asked.

"Let's just say this: she gave me something to think about."

Luke knew that speaking to the surgeon had been a kind of test. It was the first of the medical explanations (in this case a nonexplanation) he'd be given and would need to recite back over the next days or weeks. This was what getting sick was: learning a language he didn't want to know. Luke was trying not to have an interest in things he wished he'd never heard of. The surgical preview made Luke think about threat and probabilities.

He hadn't known he was terrified, Luke told me on the phone, until he ran a stop sign, never having noticed it, and nearly got sideswiped; until he cried at a movie about a boy losing his father; until he asked for a second scotch at dinner out with friends. He obsessed with minutiae: Where was the key to the safety deposit box? Who took his passport out of its drawer? Luke was in the sway of fear. Although he couldn't quite tell me exactly what he was afraid of, couldn't put into words the shape of his fear, terror moved Luke like a pendulum, swinging his heightened attention between alarm and weariness. He didn't think very much about his bump, he said; then he couldn't stop thinking about it. Terror signals an awareness of exposure, of how close one is to being shattered. There is such a thing as reasonable terror, but Luke found that he could not move on, could not acknowledge and overrule his terror.

Luke called me because he was sure that if he talked to his surgeon about his fear, she would be unmoved. Surgeons think surgery is pretty basic. They do it every day, there's not much suspense in it for them, and they're reasonably certain of the range of outcomes. They speak casually, incidentally, as if a patient's particular situation has occurred a thousand times.

Why can't patients tell their doctors about the depths of their fears? Patients put up a front and hide their full distress. It is no small thing to admit to terror. It reveals the most personal mechanics of your thinking, the connections you draw, the conclusions you reach. To disclose your terror is to warn another person that you may need to break the contract of propriety that generally presides not only during daily life but during medical life. The revelation of terror is an admission that disruption and incoherence now rule, and it is a disclosure made to a relative stranger, your doctor, who you can only hope will welcome such honesty. But such a revelation assumes your doctor can and will understand your admission in all its shadings, its context and weight. In most cases, you are telling everything to someone who knows nothing about you except the barest medical facts. It is the opposite of all you understand of how human relationships should progress: as a patient in the grip of terror, you are forced to skip the preliminary steps that build intimacy. You usually reveal your vulnerabilities only when you trust. Fear, you are trying to warn your doctor, explains why you may not be listening, why eye contact may be difficult. It is an acknowledgment that there may be hallucinatory moments beyond reason when decisions are stalled, when making sense of things is impossible.

Luke was convinced that his surgeon wouldn't even understand that terror was part of it. By the time he called me, he had reached the point of having little interest in understanding the details of his condition. All he wanted to know was: will I die? Anything more—will it come back?—was irrelevant. Any more data was too sophisticated. Facts were useless and fugitive to Luke. He had shrunk; what was left was self-pity and inward trembling.

Patients arrive at my office and expect the physical surroundings, like their care, to be neat, calibrated, and exact, and then they see the first disturbing detail of chaos, a water stain on the ceiling, a crack in the wall, a broken window, a blister of paint, and suddenly they're no longer sure what to expect. Their vague anxiety becomes clearer: a hollow uneasiness turns to fear. It's natural in any doctor's office to think: *Something terrible is going to happen.* I could imagine the neurosurgeon's office with its gray, fist-sized model of the cerebrum, its charts of the red and blue blood vessels on the underside of the brain, and Luke's terror slipping out of control.

"You're not going to die from either the surgery or whatever your lump is," I said firmly into the phone. Luke's predicament was that he believed two things at once: nothing was wrong with him, and he was doomed anyway. As I tried to reassure him, it occurred to me that if rationality is sharp, its points discrete, then fear is like a tide of dark, sticky sludge, covering everything but impossible to hold or get rid of.

Luke brought his entire family—his wife and two kids—into my office a week later while he had his preoperative electrocardiogram. He had agreed to a surgical date three weeks hence, but he wanted to hear my opinion about the option of doing nothing at all. It was early July, hot and humid. I had pulled the shade to block the warmth radiating off the window. In loose orange shorts and a gray T-shirt, Luke was sweating, his upper lip glistening. As I attached the electrodes to his wrists and ankles, Luke asked, "If it's true that the odds are in my favor, can't I safely wait on this operation? At least for a while?"

His reluctance was natural, but what was he afraid of exactly? The surgeon had told him that ninety-nine chances out of a hundred, his bump would turn out to be a nuisance but not life-threatening, not malignant cancer. Rational details calmed him, but primitive thoughts took over. His bump was not on his arm, for instance, but on his head, the home of four of the five senses. His consciousness resided a few inches below. The surgeon would be working near his brain, and Luke, a pension fund manager, "lived" in his head.

"I hope she puts the lid back when she's done," Luke said when the cardiogram was done and he was back in the chair beside my desk, his left foot tapping madly. In his lighter moods, his visual image of the burr hole to be drilled around the lump was of a kind of ice-fishing. I was pleased that he hadn't lost his ability to laugh. His wife, more aggressively, said the location of the surgery was like "catching a bullet between the eyes." Luke's older son thought the technology and the fast-drying cement to be used to close the hole would be "cool." His younger son asked two questions without waiting for a reply to either: "Is it dangerous? Will you get Jell-O?" Luke was worried about the surgeon poking around near his brain, the likely postoperative pain, the intra-operative complications, and the expected weeks of recovery. But as I went over each of these aspects of his care, assuring Luke of his surgeon's skills and fine judgment, he hunched in his chair, telling me with his drumming unease that he was crowded by some other anxiety as well.

I suspected he was worried about something he hadn't told me about, and I thought I knew what it was. "Are you anxious about going into the hospital?"

"He's never spent a night in one," Sheryl answered.

Watching Luke, something else occurred to me: when you are well, if you have thought of death at all, it is as a natural force, wild, unpredictable, a lightning strike on a sunny day. When you are ill, death is the storm you see approaching. Once you have seen this (and early on during any illness almost all patients do), you don't feel quite the same again, and no one seems to speak to you the same. The change is tiny, but huge. The terror of death had stolen Luke's future.

Terror is a dry run for the dramas of illness, of mortality. Illness reminds you of your limitations every second, and *life's* limitations in particular. I suspected that Luke's general awareness that "all men must die" had been replaced by the specific awareness "I must die." With the arrival of terror, the dying process has already begun. Terrified, you have begun to understand that your life will end, that you *can* die.

When he called the next day, Luke informed me that he was probably not going to have the surgery. I was surprised. I took it for granted that when a surgeon suggests an operation, the patient complies, even when the surgery is elective. "Well, if you haven't canceled the date, you still have a few weeks to decide," I said. I presumed that Luke was disturbed by the image of himself horizontal, exposed, and that he had Turgenev's sharp fear, the sense of a surgeon's scalpel "like a knife cutting into a banana." Even with a growing mass on his forehead, if a terrified patient like Luke feels well, his inclination is to go home and risk whatever might happen. Luke had convinced himself that the risk in waiting was low, or that there was no risk at all.

The mind is more persuasive than a doctor's recommendations. Any patient can convince himself that he can wait out illness. It is reasonable not to trust the doctor's worst-case scenario. The patient goes over the odds: *I've lived this long, so why gamble with interventions?* The doctor's answers to questions are useful, but beside the point. The patient asks himself: *Why risk a procedure, the evaluable complications, the unanticipated troubles?* For Luke, doing nothing made him feel *less* passive, *less* of a victim. I told Luke that surgery is like a prizefight: violence followed by a handshake. But he would have none of my clever analogies. We spoke on the phone for a few minutes more, and then Luke made an admission. I had missed the point: he was not afraid of the scalpel or the hospital, but he was deathly afraid of the anesthesia.

Terror is made worse by being alone, but being alone seems to be the only condition a patient wants when he is scared. Sheryl called me a few days later. She was worried about her husband, who had missed the last three days of work. He had become caught up in himself. He kept touching his lump absently. Because he didn't want his friends to think of him as terrified, to see him that way, he had backed out of his regular Friday night poker game. If Luke spent time with his friends, he was afraid his terror would be confirmed, that there was a good reason to be terrified. He wanted people to tell him that his diagnosis might not be so bad, but he knew they might not say this. He didn't want to be observed too closely. Alone, he could suffer without embarrassment or constraint, without fear of weighing down others or wearing out their sympathy. Keeping to himself forestalled the disgrace of revealing his terror. Sheryl understood that Luke could not be humiliated if no one was around, so he made himself unavailable to be helped. She didn't want her friends to think her husband was rude, but he didn't hear people when they spoke to him, and if he did, he didn't really get what they were saying or remember it after they were gone. "It's not easy to tell the truth about how you feel when you want so badly to feel differently," I said to Luke's wife.

Of course, being alone, a patient can't get out of his own way. Terror is its own obsession. It is a form of madness that I can hear in a patient's voice. Terror is the mania of trying to get your mind off your mind. Initially a patient can't accept that he might be sick, that surgery could uncover something disastrous. Ill, a patient spins his words over and over until they are suddenly meaningless as a way of *not getting used* to the idea of illness, of closing off from it.

I wanted to tell Luke: I know what you're thinking. You are trying to outsmart terror by turning the fact of your skull mass into a game. For Luke, being scared seemed to be caused mostly by thinking. If he could find a rhythm that was so busy that it excluded thinking, he'd be all right. But not thinking seemed wrong too. If a patient is terrified, not-thinking is really *pretend* not-thinking. It is an effort to keep fear at bay.

The terrified patient is reduced to a single mental command: make time pass until you are well and safe again. To follow this command, he needs to dissociate himself from his known surroundings. Here, fear is beneficial to survival, the sense that there is a threat nearby. Limitless fear, though, can be destructive. When terror assumes catastrophic proportions, it distorts one's vision. The patient makes himself go blind, and everything he knows becomes invisible because these objects—the weave of his wife's blue silk blouse, the shape of a perfume bottle—have associations and memories attached, and he is incapable of looking at the past, or at any image other than what awaits him. He steps away from so much of what fills life because it feels like it's already gone; he feels, in a sense, that he is never coming back and will never live with these objects and memories again. He is about to be sent someplace that he can't see, that stretches off into nothing. He closes his eyes, attempting to rest, but the unrest that follows offers no obvious benefit.

Luke came to my office, alone, a week later, looking miserable. "Every time I pass a mirror I worry that I'm making the wrong decision." He was still decided against surgery.

"What does Sheryl say?" I figured she had lost her patience with Luke's defeating fear.

But he was so afraid of surgery that he had entirely stopped speaking to his wife about it. For the most part, the possibility of agreeing to anesthesia didn't even exist. If he mentioned it at all, he sounded hysterical, pessimistic, crazy, overdramatic, he said. I realized that he was coming into my office because he had no one else to talk to. Luke possessed nothing but terror; he was experiencing nothing beyond a mere sense of existing. He couldn't pin down his terror, and it preoccupied and troubled him: it was pinning *him* down, and isolating him.

A patient heading toward surgery is enclosed in a pocket of timelessness and anonymity. He has absurd thoughts. All the people who went before him to surgery, where are they now? It feels as if his operation is infinitely far off, but also as if he is in a line pushing forward to go through the tollbooth. He fumbles for change, the worry presses until time disappears, until every moment seems worthless, unimportant. The next moment he's suddenly aware that it's his turn to pay.

"It's worst at night," Luke said. I had heard the same thing just the month before from Mrs. R., who was awaiting bypass surgery. At night she was filled with resentment, anger, and resignation over what had befallen her; she suffered or plotted a way out. But she saw none. "Sheryl falls asleep instantaneously," Luke continued, sounding frustrated, "and I'm too tense to close my eyes." Sometimes late at night, when he was the only one awake, Luke liked to imagine he was keeping guard. He would be the first to hear the steps of the intruder; he would announce the first warning; he would be ready to move instantly. Somehow this idea that he was alert to the dangerous world gave him comfort. Oddly, terror made him immune too. After all, what else could happen to him now?

But the night before he had felt his heart "unraveling." "You know how sometimes you give a tug on a spool of thread and it spins to the floor?" he asked. I imagined his pulse in frenzied flight, hurtling and heedless, and I knew that *Luke* was unraveling. During the episode, he felt lightheaded, tight-chested, frantic. I had had an attack like this myself soon after the drive when Richard delivered his news to me. Panicked air is cold and thin and hard to breathe.

Terror has a physical aspect. Terror made Mrs. R., a thin woman in her late fifties, feel thick and substantial. She'd been targeted, and danger couldn't miss her. Some mornings she woke up shaking. Other mornings she ground her teeth to get through to noon. Her swallowing sounded loud and distinct. She reported that the smell of her sweat, the scent of terror, was vaguely repellent during the week she waited for her heart operation.

"What should I do if it happens again?" Luke asked, not caring to discuss the likely cause.

"What *did* you do?" I answered.

"I got up and tried to walk it off, and slowly the feeling went away."

"Did you wake your wife?"

Luke didn't answer me for a moment. Feeling guilty about the anguish he had been causing her, he hadn't woken her. Terror controlled his mind, his mood, and now how he treated his loved ones. "I'd already told her three times that day that I was sorry for my behavior, my absence. That seemed enough. And now this."

That Sheryl had appreciated his apology, his recognition of her concern, suggested that he wasn't completely caught up in his own mental loops. "I bet she would have wanted to be woken if she knew how upset you were, how alone you felt," I said.

"I actually appreciated her sleeping like that," he corrected me. "She's the only one with a normal sleep cycle in my house anymore."

He was calmed by seeing his wife resting. He wanted to protect her sleep because he got vicarious comfort from it. "Sometimes our minds are forced to follow our bodies," I told him, but if he wanted, I could give him an anti-anxiety medication.

"The only part I'm afraid of is the anesthesia. I have this

mental image of being abandoned. It's just freaking me out. I don't know why, but I just can't do it."

Near sleep, terror becomes a form of self-hypnosis. The patient tries to empty his head, accept his fate. Terror is a way of taking himself so deep into his imagination that he feels outside himself. The brutish truth must be a dream as he concentrates on his humming fear.

At first, I had thought Luke's fear was vague and impressionistic, a mistrust of the unknown, but now I knew that it was precise and intense and involved the anesthesia that was necessary for the surgery.

Luke was sleepless. Anesthesia is a form of sleep. Some patients think of anesthesia as taking a few hours away from the normal flow of hours, a vacation. Others think of surgical anesthesia as death, as time lost. Being "put to sleep" has bad connotations: *If I can't be awake through all of this,* patients say to themselves, *things must be pretty bad. They don't want me to "see" what's going on.*

Anesthesia depends on the competence of others: the patient must completely give over control. Going under signals the end of vigilance and self-protection, the fundamentals of patienthood. Patients must agree to trust blindly, to let others take over this concern of theirs. This is a gamble. It holds the possibility of not waking up, of being duped, of relying on the wrong people. Anesthesia is closer to death than sleep because it is characterized by the absolute cessation of physical movement. To Luke's mind, anesthesia was a state of mortal loneliness, an empty mental chamber. Anesthesia might not have been death, but it was the suspension of life. And if he was lucky enough to awaken, there was a second, different terror. He asked, "What if I survive and my brain does not?"

There is no predicting what will terrify a person.

Chained up in a patient's head, terror causes untold physical wear and tear by expressing itself forcibly. The pressure Luke felt to make the right decision made his head seem like it was ready to explode, and this condition seemed a permanent, unyielding state. There was no repose. He was angry because he couldn't free himself. Terror weakened him, and he wished it would just go away, or at least dissipate. But it didn't. And physical exhaustion arrived with rage, the result of frustration and sleeplessness. He railed against the world because the world had produced this fix he was in: this bump, this undiagnosed illness, this need for surgery, this fear that sickened. He didn't know how to imagine a promising future beyond anesthesia. He could not distinguish it confidently. It was there, but it held only the sad and unwanted.

Neither the patient nor the doctor has a script. I've learned over the years that there are cues to a medical visit but no plot. During my first years in practice, I thought I was a competent enough performer for this improvisation, but each terrified patient cast doubt on the premises of my medical training—that my answers mattered, that I held crucial information—because each of these patients was not listening. Nothing I said quite made sense.

Undefended, innocent, betrayed by his own body, the terrified patient has lost the thread of the narrative of his life. To him, the next hours will merely be another series of temporary arrangements—walk across the room, try to eat dinner, raise arms above head to pull off your shirt. All doctors are taught to keep a safe remove, but the patient overwhelmed by fear draws us in. I have come to understand that it is acceptable to try hardest with the most apprehensive of patients, to offer reports about former patients who've made it through the painful diagnostic testing, the chemotherapy, the surgery. Our job as doctors is to try to make the larger narrative intelligible again.

I accepted Luke's fear, but I wasn't sure how best to help him. What usually terrified patients—lab results, how they felt, the wait for diagnosis, the rush to treatment, the body's response, the seeming loss of choice—did not apply to Luke. He was terrified of not being a spectator to his own drama, his two hours on the surgical stage.

Luke had given up the possibility of being comforted by me or anyone else. He didn't want to hear, "You're going to be fine." If I had said to Luke, "You're being very brave," he would have answered, "Go to hell." Patients do not think of themselves as brave because even as they object, somewhere inside they are accepting that *this is the way things must be*. Courage, in Luke's case, meant not scaring others. He wanted to prevent those he loved from suffering. At the same time, he didn't want to do the work of making his siblings or kids feel good. He didn't have the energy to expend. He didn't want to talk about the option of discussing his fears with friends, whom he saw not as supporters but as a network of rumormongers.

He no longer felt he had to apologize for showing emotions other than laughter. He started to pace. He pounded the examining table. He yelled at me for the hole in my left shoe. I didn't want Luke to hide his strain and exhaustion from me. Illness and terror are about fighting, about not giving in but wanting to, wanting to very badly. Anyone who perceives himself as close to death deserves to rail. I didn't mind that rage accented Luke's every question of me. I wanted to hear his full itinerary of despair. Terror wasn't a temporary outburst but an ongoing emotion for Luke. His acting-out reminded me of Richard, who had no interest in good manners, in keeping a good attitude, in being positive, in smiling. "If my body is out of control, why shouldn't I be?" Luke asked.

To lose control is to fail. Maintaining control is good manners as well as moral duty. A patient's terror is out of control because his body is out of control. He understands that surgery is indeed about giving over control to two people, a surgeon and an anesthesiologist, who have given themselves over wholly to what they are doing, and what they are doing would be *him*. Surgery is pure control: single-minded, austere, absorbing, clean, a site of order. But the patient asks: Are they tough enough? Will they protect me? Can they get me out of this? Do they care enough? The patient likes the idea that the surgeon will regain control of his runaway body, but he still doesn't want to undergo surgery.

Patients don't know if I've ever experienced the kind of dread they're trying to describe, and often they think it's not worth explaining because I will be part of their life for only a few minutes, or a week, or, worst case, a few months. At a certain point, they recognize that it isn't realistic to expect my appreciation of their situation, their crisis, ever to be complete. I will not be operated on alongside them; they go off to surgery alone. Illness is a barrier, invisible but everlasting. But they continue to ask questions, in part to hang on to control, in part to receive some minimal comfort from me, in part to discover what they need to feel better.

Some of the answers I gave to Luke's recycled questions were useful, but many weren't really answers at all. When he asked, "What if I survive and my brain does not?" I responded with chances and odds so small they were impossible to summon or calculate. We went over it again and again. I could see his mind buckle and flex, assert and cringe. But by asking questions, patients feel less powerless, and by answering them, doctors try to alleviate terror. I know there is no drastic remedy for terror, but what patients want and need badly is, as Reynolds Price wrote, "the frank exchange of decent concern." I try never to turn away from conversation. I give looks of mild encouragement. I offer words, nothing altogether convincing, but at least amulets of hope. As Luke was leaving my office that day, I said, "I will be at your eightieth birthday party."

Maybe he would be eighty years old, but not this week, with his surgical date approaching. If anything, he probably only wanted to be praised for expressing terror so openly as he paced and pounded.

Here's what I have come to understand after fifteen years of work: patients should not be expected to be courageous with their doctors.

In the midst of his pacing, Luke pulled his cell phone from his pocket and called Sheryl. "My wife wants to talk to you," he said, handing me the tiny silver rectangle.

"His lips are white, and he keeps touching them," she reported, panicked a little herself by his new habit. I knew that Luke's terror had burst through and covered his life. "Things that really matter to him—his woodwork in the basement, his fantasy football league—he's been completely ignoring." I thought of Luke trying to remember a time when fear seemed natural, part of life, rather than cataclysmic. His enthusiasms had drained away. There seemed no end to his fright. No day was short enough. Touching his lips was a way of trying to keep it all contained, I tried to reassure Sheryl. "You have to help him. I can't," she cried.

I wanted to get Luke thinking not only of the surgery itself but beyond it. I never expect to return my patients to comfort *before* they've gone through surgery or the worst of illness, but I can remind them of the pleasures in the world, and of what they will return to when the terror passes.

I handed the phone back to Luke, and it was Sheryl who finally convinced him that he had to have surgery, risky or not, because *she* had to know if his lump was serious or not, even if he was willing to live with the unknown. Luke lowered himself into a chair. I watched as he drew tantalizingly close to a solution for that problem that weeks before had seemed impossible. Speaking with his wife, he understood that he could not escape this operation, and the muteness of fear became very much like peace. He stopped asking himself, *Do I really have it in me to get through this?* He understood that all the *theories* about what was wrong with him didn't matter. What mattered was the doctor completing the procedure, the final pathology report, and then doing the work of recovery.

Terror is the beginning of the end of the illusion that illness isn't that bad. It precipitates the breakdown of keeping up *appearances*. I was used to seeing patients make every effort to be optimistic, cheerful. I am always upset when I recognize that a patient is terrified: their fear breaks my medical facade, and I too feel vulnerable. Richard, before he was sick, had never had an interest in appearing positive and good-humored. He had no interest in being stoical. (After he accidentally smashed his thumbnail with a hammer, he begged me for narcotics.) So I shouldn't have expected Richard to conceal how illness was affecting him. But I did hope he would, because his cancer terrified *me*. Luke's fear had been so high-pitched, so contagious, that I too had become worried about his surgery.

I stopped in at the hospital admitting area the morning of his operation. Sheryl was holding his hand in a small room with a window that faced a garden filled with cool yellow light. Luke told me that the expression "frightened to death" passed through his mind and that he hoped it wasn't true. Then he set his clothes and his terror aside, lay down on the green-sheeted stretcher, and proceeded.

Oliver Sacks has pointed out that we have the word "sickening" to express the arrival of illness but that, interestingly, there is no word "healthening" to mark the beginning of improvement. The opposite of sickening is *recovery*, a word that suggests something was lost during illness. Lost is the sensation of being well, a perception that is unnoticed until it vanishes. During the hours of Luke's operation, I realized: the *patient* has been lost, he has been exiled, he has been gone. When he has recovered, he will have returned. He will be renewed and restored. At the end of illness, the patient reestablishes residency in the land of the healthy, even if this return is temporary and accompanied by a recognition of losses.

I visited Luke as soon as he was moved from the recovery area to his hospital room. During the two-hour operation, his surgeon removed a bone-based hemangioma, a benign tangle of blood vessels that had grown anomalously, bending his skull, but was unlikely to return on his forehead or anywhere else. He required no further treatment. "I made it," Luke said. He seemed almost surprised to be done.

"Not a problem," I answered.

"For you." He smiled, remembering how to be amiable again. He was impatient to get up; he had never thought of himself as delicate or fragile. He swung his legs to the side of the bed, and I helped him stand and turn toward the bathroom. He was dizzy, and his forehead was wrapped tightly with a pressurized Ace bandage. "It feels as if my brain is being squeezed out my ears," he said, wincing. Luke's terror, which was unendurable only the day before, had disappeared, abruptly and completely. The demands of terror had seemed impossible to meet, and now it was as if they had never existed. Or as if Luke had entered a period of thoughtlessness.

The resolution of acute, temporary illness has a certain simplicity that discourages too much curious post hoc speculation on the patient's part. Still, I wondered how Luke now thought of himself. The successful completion of surgery and the immediacy of incisional pain had, at least temporarily, displaced his fear, so I asked him how he had gotten through the terror of the last weeks. "I let my knees shake. They would just start rattling, and I'd let it happen. Nothing like a little unsteadiness to remind me of the old feeling of omnipotence," he offered, smiling, returning partway toward his old self, the old country. "Do you think there was a better way for me to prepare?" I saw him wonder what his fear had even been about. If I hadn't reminded him, he would have forgotten it altogether.

His eyebrows were squashed toward his eyes by his head wrap; any expression on his face now involved a wrinkled nose and curled lips. The bandage was meant to keep down swelling, but it made the rest of his face look swollen. Luke had pride, and his recovery would prove his durability. But he was still in a subdued state, despite his bluff demeanor. Moments of trembling were left over from the acute time. Never again would Luke be totally and thoughtlessly sure of his body. Terror had brought him to the edge of himself, allowed him to measure what he was capable of, and he now doubted himself, despite his knee-shivering strategy. The wrap was squeezing tears from Luke's eyes.

The initial happiness of recovery is that of the escaped convict. There is a sense of jubilation, of having outrun pursuit, but the escape is not quite free and clear. The liberated prisoner still thinks of himself in terms of his time in jail, and the recovering patient lives in reference to illness. Recovery includes a constant sense of watching and comparing. How am I compared to *before*? Time and goals have become confusing. Patients think of the days, weeks, years before they were ill as "real life." *Real* life, then, is *past* life, and patients want to get back to it. But one's past life, the regular life of endless good health, its buffered permanence and protected space, is illusory, the recovering patient now recognizes.

Yet if the patient can say, "I feel like myself again," she can almost convince herself that illness never existed. Lucy Grealy, the poet recovering from a rare tumor of the jaw, wrote: "It's funny, but I'm always forgetting I had cancer. It seems like a different person that happened to. . . . I wore a red dress. I was in a state of despair. It wasn't me at all." Illness is part of real life, but it feels separate, and during recovery patients want to keep it separate so they will be exempt from its return, its being part of everyday life.

Recovery is not always a straight path. It is often fractured and confused. The sternal wound became infected after Mrs. R.'s bypass surgery. Breast cancer returns inches from its primary location before chemotherapy forces it into remission. Multiple sclerosis relapses despite suppressive therapy. The victory over illness seems less impressive than we might have thought. It seems almost haphazard. If illness is a matter of chance, so is recovery. The patient understands, perhaps for the first time, that health is a condition, just as illness is. It is a state of effectiveness, at least bodily effectiveness. When a doctor says, "You are going to recover," the patient has been given a suit of summer linen to replace the hair shirt of illness, but the outfit can change again. He has received permission to improve, but he may be reluctant to accept it. Melancholy arrives as the patient accepts that he is not invincible; if this has happened, then anything can happen. It is time to make a clean start, except there are no clean starts anymore.

Within hours after sinus surgery, Richard was also being urged to stand. But he was in no rush. He dallied in bed, pulling the sheets and thin blue blanket up to his neck, closing his eyes so he could ignore any suggestion he didn't want to hear. Recovery does have its blissful aspect. There is a lightness that comes after misery. Rabbit Angstrom, in John Updike's *Rabbit at Rest*, describes the pleasure of recovering from his heart attack as "being in the hands of others, of being the blind, pained, focal point of a world of concern and expertise. . . . Sinking, he perceived the world around him as gaseous and rising, the grave and affectionate faces of paramedics and doctors and nurses released by his emergency like a cloud of holiday balloons. His many burdens have been lifted away in this light-drenched hospital, this businesslike emporium where miracles are common if not cheap."

Once the indiscriminately strong assaults of fear and pain are past, patients sometimes feel invulnerable and immortal. Richard experienced a sense of pride, the confident, well-off feeling one gets after performing a great athletic feat. The hospital room was small and unadorned except for a vase of flowers. One side of Richard's face was gauzed and taped that first postoperative day, but he smiled with benevolence and amusement. In his story "The Cure," John Cheever describes a "feeling that was far from happy but also with that renewal of self-respect, or nerve, that seems to be the reward for accepting a painful truth." The painful truth of his biopsy report was still unknown during those hours, so Richard was savoring an element of triumph.

A patient allows himself a certain selfishness when he senses it will be tolerated. Richard loved the constant action around his hospital bed, and he liked being taken care of. When nothing was being asked of him, he enjoyed the pampered, cherished status. He relished the attention, the ability to order people around and ask for things, the call button under his thumb, his family hovering. He didn't seem to notice the hospital smell of thin coffee and filmy detergent. Tired, he was reminded of his incapacity and inadequacy, but in the hospital I think he also felt relieved, a feeling that included a sense of reprieve from the callous, careless, demanding art world that was likely to reject his newest work as soon as he returned home. For the moment, he had no responsibility, to anyone or anything. Everything—and most importantly, his body—was being looked after by somebody else. As Updike writes: "Vast portions of the world are shorn off, suddenly ignorable. You simply become a piece of physical luggage to be delivered into the hands of others."

I sat beside Richard's bed and, trying to inspire him, read aloud a passage from a book that tells the story of a soldier who, after being accidentally shot in the chest during a training exercise, is operated on for five hours and that same night wakes and says, "I am not the norm. I'm ready to get out of here and I'm ready to prove it to you." He has the doctors pull the tubes out of his arm, hops out of bed, and does fifty push-ups. They let him go home. When I stopped reading, Richard said, "I used to do two hundred and fifty push-ups a day. *Those* were the days." Enjoying his pain pills, he was in no hurry to get going or cut short his convalescence. He was not yet ready for the grueling mule-work of recovery. He had no interest in going outside to see what kind of day it was.

"Erectness is moral, existential, no less than physical," Oliver Sacks wrote in his memoir of recovery from serious nerve damage in his leg. The posture of a patient—recumbent, confined—is one of physical weakness and moral passivity, of a man reduced, dependent, living within the limits of his bed. Illness is a state of passivity, and humbled, profound passivity may, for a time, be the only proper attitude when you are sick. In bed, ill, things are literally *out of reach,* the phone, the pen, the television clicker. You can't get them, and at first, you don't care. You don't want to overreach. Drowsy and in pain, you seem to have few choices left; none of them are interesting, and all are disturbing.

The recovering patient is never sure how to act. His head teems with calculations of what can and can't be done. He's not sure what to ignore, what to get angry at, what to worry about, what to be reassured by. Decisions are difficult and fraught with hesitation. The connection to the ordinary world is broken, but not entirely. Yet the patient is filled with a pure and uncomplicated longing for health. He is intoxicated by possibility after powerlessness. A patient always holds some secret notion of promised health. He listens carefully for good news, which will confirm his secret wish, and when he gets it, he never forgets it. Still, sometimes he lies in bed an extra day or two, reminded of what he was like when he was powerful.

Sacks knew he was recovering when he reached a distraction point and realized that what the nurses said was true: he was a deserter from the army of the upright. Bed-bound for too long, the patient inevitably feels condemned, captured, disallowed from participating, and ashamed. Standing is often an ill person's first act of real recovery, his first act of participation, an experience that is hardly spectacular, since he has done it without thought, perfectly and easily, countless times before in his life. Standing and walking are escapes from the gravitational pull of the woe and resignation that is illness.

In bed, the patient thinks, *Is this what I will have to put up with for the rest of my life?* Walking is the return of *doing*, of going, of regaining control of the physically possible. It is a metaphorical "getting on with it." The patient is proceeding—he is "taking the first step." Richard's first steps were clumsy, not fluid or free. Dragging tubes, he was still thinking out each move and wondering, *How did everything get so complicated? Can't it be easy?* Richard's surgeon finally insisted that he walk, and he held on to my brother-in-law the afternoon of the second postoperative day; holding on, of course, in its second sense, signifies survival. Richard fought him, wanting comfort and serenity, an atmosphere of comprehension and sympathy, while his doctor wanted movement. Richard's slowness was also a matter of obstinacy and attitude. He presented his condition in as bad a light as possible, for the worse he made it seem, the more license he had to be annoyed and angry that he was being pushed, that his duress was not being taken seriously enough.

Patients like sympathy but don't trust it. They've already been betrayed once. "That wasn't so bad, was it?" I once heard a surgeon say to a patient during his first postoperative visit, and I thought it was the most hateful question. Although offered sympathetically, it had no delicacy. It implied that what the patient had gone through was nothing; with the suggestion in his question that his own life was simpler than his patient's at that moment, the surgeon exposed himself as a fake, someone the patient had only mistakenly thought he could trust to understand what was significant. Escaping from doctors, even if they've saved you, is another impetus for recovery. All patients believe that if they can get away from doctors, their lives will improve.

Do patients ever trust recovery? It involves more than trusting doctors who announce: you will get better. It is a matter of time and discovery. The patient is starting life over. Recovery is fraught with worry: Is today's improvement a false sign? Has the sentence already been passed? The patient may have recovered part of his future (the part immediately beyond illness, which is the only piece desired when one is sick), but not all of it, and there is concern about what will happen next.

"My body started orienting itself toward home, feeling stronger and more bored every day," Lucy Grealy wrote in *Autobiography of a Face*. Even at home, recovery is purely anticipatory, and hope is always theoretical. It is never completely clear what there is to be hopeful about. To get one's hopes up and have them dashed is humiliating. Before one has taken the first step, it is not even certain that the necessary mechanisms of recovery will be available.

But unlike hope, its philosophical sister, recovery is practical, dragging along the body with its needs and substance. In some ways, we expect recovery to reflect illness: when illness is acute, abrupt—strep throat, for instance—we expect recovery to be the same. Richard's original symptoms had evolved gradually and were barely detectable until their cumulative effect made him notice the change, the numbness along his palate; then he finally sought help. His lounging in the hospital a day longer than his doctor wanted was the first sign that he expected his recovery to be prolonged as well.

Recovery is a project, a daily effort. It does not arrive in a single, decisive parcel. The progress of recovery, slow over weeks, seems paltry. More than anything, recovery requires tenaciousness, whether the illness is one that relents but never completely disappears (like multiple sclerosis) or one for which surgery is definitive but disruptive, as with Luke's hemangioma. In either case, health must be won and then controlled, and often won again. There are intervals when the patient is ready to give up, and others when he believes again—despite the evidence—that wellness can be a permanent condition. There are times when he decides to stop feeling sick, but his body overtakes him and he continues to feel sick. He wonders how low he can go. He decides and undecides; he surges and reverses, rallies and relapses. Illness was freedom taken away—he couldn't do what he wanted. Now he wants what he wants.

Unlike illness, recovery has an agenda, and one that is patient-controlled. There are physical skills to learn—how to use a wheelchair, or self-catheterize a bladder, or walk without a cane, or fully extend a knee, or reach a high shelf, or, in Richard's case, breathe deeply through the nose—and emotional ones, such as how to accept care, or laugh, or enjoy food again. Mastery returns and luck is restored. A sense of triumph floods in.

Recovery is certainly not all pleasant. Many mornings bring on impatience, irritability, and an increasingly bad temper. Activities seem either urgent or makeshift. Everything is an inconvenience, an annoyance, a hindrance—the oven heats too slowly, rain interrupts the promised sunshine. The patient is a walking grievance, sullen and unpredictable. Pissed off, he overreacts, monitoring all in his life that he's still missing. Illness provided constant fuel for his bad temper if he was of that disposition, ready to shake his fist at the world; recovery provides a different fuel as he tells the truth about what happened—a not-so-simple act of self-assertion—and watches how people treat him. He bolts toward the clear territory of health, toward home, but the escape from illness feels precarious.

More and more, gratitude supervenes. Crowded, the anxiety of illness is pushed away. The patient's concerns shift and generosity returns. He has an interest again in admiring how well his body works. He is willing to skip over the naive, large questions—will I die?—when speaking with doctors so that he can hear about the function of arteries, nerves, joints, immune system. The patient's curiosity about his doctor had been directed toward only one question—can you fix me?—but now he can appreciate the doctor's work. For the first time, he studies his doctor's face. Before, he hardly knew what his doctor looked like; doctors were functionaries. Now he thinks of his doctor watching and approving. He can actually visualize his doctor's lips saying, "Good. That's right. You're doing well.

Keep it up." Illness recedes, but health does not immediately step up and take its place.

Recovery takes on a different kind of touristic mind-set, one whose chief characteristic is receptivity. The patient is intrigued by the new again, and details are interesting and valuable, possibly useful later on. He looks around at a place that it suddenly seems he's never been to—his bedroom, for instance—even though he has looked at it every day for a decade. He collects ideas for the future (*maybe I should move the bureau over there*). His attention is reawakened. He may have revised certain notions of the good life, but his enthusiasms have returned. He is glad they are back; he is glad to be back.

To be fully recovered is to be returned to yourself. You can again be in contact with your emotions, your longings, the ideas that are important to you. You are responsive to the world again. It is the time when you prefer to be deliciously nude, to smell almonds and mustard, to remember the lyrics of songs. Once again you love the external world, the new knobs on the kitchen cabinets, the shape of a shoulder. Recovering, you feel hungry, once again craving the good things and what Sacks calls "the certainty, the comfort, the assurance, of being safe." At night, you experience again the sleep of trust.

Nonetheless, patients remain superstitious. No longer a visitor to the state of illness, the patient does not want to look behind, having crossed the border back to health. There is a sense of reprieve; what seemed entrenched (illness) is gone. A sense of innocence and optimism revives. Skepticism and self-doubt fade. Filled with a sense of freedom and expansiveness, the recovering patient sets aside prudence. He thinks: *I was inadequate and inferior, but no longer.* Expectations are high. It is a relief to revert again to having recognizable, easily identifiable feelings. There is pleasure to be had, and patients are ready to give themselves over, to be absorbed by it. If illness is, as Zora Neale Hurston once wrote, the "boiled down juice of human living," recovery is the juice reconstituted but also diluted.

The primitive exuberance of putting up a good fight has been activated. Profound and passionate feelings are aroused in pursuit of health. Sometimes a patient presents a false energy so as to conceal the truth from himself and others: he is not quite back. He can *be* well if he *acts* well, the patient says to himself. It's why he tries to do too much the morning after the flu's peak fever: to prove nothing has changed. He might not be better than what he was before illness, but he's not worse either.

Knowledge of illness passing gives the patient a sense of power. He's traversed a foreign world. He knows its crevasses, its smells, its black and cruel zones, its great expanses and chasmal spaces. If he has to return to the land of illness, the voyage might be less drastic because he's seasoned. He's developed an admirable toughness and strength through private, inner travel, although this doesn't mean he ever wants to go back. He believes that he now has a knowledge few people have, forgetting that many people he knows have been ill.

Recovery brings the sense that something good is about to

happen. While I was treating Luke, I had dinner with a man who traveled extensively, leaving the country on an adventure every six weeks. To me, his journeys sounded exhausting, but he viewed travel as a challenge, a form of testing himself. He enjoyed navigating the roads, battling other drivers, finding places to eat and to lodge, and trying to communicate. He never stayed still on the road and took it on with a valiant intensity, vowing to learn something he could use on his next trip.

When I saw Luke a month after surgery, he told me that he was exhausted. The exhaustion came from the sudden relinquishment of terror. Yet at the same time, he felt lost in an unexpected way. I was surprised when Luke told me that he felt particularly inquisitive when he was ill and that feeling had disappeared. He thought about himself more acutely when he still had his bump; he noticed things, he said. He had been sensorially aware, but at the same time he had been so overwhelmed by terror, so monopolized by the implication of leaving his children fatherless and his wife a widow, that he had been unable to appreciate this alertness. Fearful, he had felt an unimaginable otherness, an unspeakable distance, a tragic radiance. Nonetheless, he told me, during the reign of terror he recognized more than ever before his likes and dislikes, his opinions: perfumes and colognes were too strong; daylight seemed pale; sounds were too faint, apparitional. Only now, well past surgery, was he able to share the profound feelings that had silently mixed with his terror. He told me that when he first arrived home and heard the sound of the latch lifting on his door, he felt *himself* unlock. He kept saying, "For days I had that sound of the latch lifting in my head."

During serious illness, many patients hold no secrets. A great part of my medical work is listening to people in the intense privacy of unburdening. They are ill, but while they are still vital, filled with avid strength, they confess. What do they require to get started? Almost nothing. I hardly need to bring them along; they speak openly of their wants, fears, needs. They expose themselves, sometimes fearing it's their last chance to explain. They feel brave doing so, spending their courage. Hearing secrets is one of the great privileges of practicing medicine.

Luke, on the other hand, had been slow to articulate his feelings of terror, his loneliness, his image of anesthesia as a form of abandonment. Now cured, he had little incentive to disguise or conceal his feelings. I became Luke's one-person historical society, but only *after* he had made it through surgery. He wanted to make another follow-up appointment with me just to be able to keep me apprised of how well he was doing. Yet he believed illness had been the perfect precondition for insight. At first, to contain his terror he attempted to learn about his body's complex internal functioning. Then, he told me, he had been driven deep into himself. He adopted a myth of redemptive inwardness. Without smirking, he now asked himself serious questions. What is the purpose of trouble?

What does one learn from sorrow? He had experienced deep feelings, suffering. He had been given the opportunity to uncover part of his character he had never revealed. He had new, visceral information about himself. He believed that nothing would ever be quite as intense as illness, nothing else would ever be quite as stark after cheating death and surviving. Difficult events had encouraged him to entertain difficult truths. He understood, more clearly than ever, what he meant to his wife and children and to others.

Having reached the far side of a possible catastrophe, Luke made a full and joyous recovery. But even in the midst of chronic illness, the writer Reynolds Price similarly compared his life before spinal surgery to his physically damaged life after, and he saw some improvements. He thought of himself as newly patient, less judgmental. But he wondered: Is this new view unjustified, lunatic bright? Is the ill person always self-deluding and self-serving? Is he simply unable to admit to his actual state of devastation? Or does it come down to "a normal animal pleasure in being at least alive"?

Even at his last appointment, Luke, like Joanna, no longer expected the ludicrous guarantee of perfect health. Safety was illusory, another betrayal certain. Luke saw his experience of illness as a strength-building. He had a deeper relationship with himself, and he was ready to set better life priorities. But it would always remain shocking how his body had betrayed him. He told me that he could easily be persuaded that the lump on his forehead had never happened.

When a patient regains the sense that death is far away, he begins keeping secrets again, even if the secret is simply that he just tasted a wonderful hamburger, that wondrous and forbidden delicacy. Not being able to eat what you want is a sad affair, and all too common during illness. When Richard and I went to Johnnie's for the first time after his surgery, he stabbed the air between us and wiped his mouth. "Don't tell your sister what I ate. It's between us." Secrets are a way to assert the self again. They speak of the future, of holding on to something private.

Looking back, we see that illness is a matter of critical moments, of long and peculiar and memorable days. Richard's memory had allowed him to omit some of the awfulness, but now, during recovery from surgery, he said he was unlikely to review the extent of his suffering. He preferred not to inspect what he had gone through but rather to respect it and leave it alone. He was one of those who believe that dwelling on one's problems only adds to them. I had imagined that Richard would use those first weeks of recovery to reconsider his life, but the opposite was the case. He didn't want to examine things too closely. Doing so seemed like an invitation for bad luck. He had lived under the artifice that good health goes on forever—the surface story we have all agreed to live by—and to consider otherwise felt too risky, a jinx.

During his few days at home awaiting the pathology report, Richard temporarily escaped not only his cancer—the memory of which he conveniently lodged in an obscure corner of his mind—but also the young doctors with their clipboards and defensiveness and poking arrogance. He had escaped the young, overworked nurses, with their hard-to-believe cheeriness, who were able to pretend that everything was normal. He had escaped the one older nurse with her distant warnings and lofty murmurings and absolute power. He was finished with the needles and X-rays, his wife's parking problems, and her full-time worry over having a husband sleeping in a hospital. He had finished with the numbing conferences and counseling, the scouring, wrecking medications, and the vibrating machines as spacious as confessionals where his organs and bones, which he had turned over to others, were revealed. He no longer had to fight the natural inclination of the mind to shut down. In his gingerbread house, Richard had returned to his toothbrush and his cats, four blocks from Johnnie's Luncheonette, where his illness was a necessary, but not defining, aspect of his existence.

The pathology report revealed that cancer remained deep in the crevices of his inner skull. There was no saying how much was left or how quickly it would grow, but it was there. My sister reported that the surgeon's gentle tone and optimism made the news manageable and negotiable and gave it a shape that could be handled so that its content lost some of its infinite abstract power. Given the tumor's location, his surgeon suggested, there was no further treatment planned, just watchful waiting.

For some, illness is an opportunity to stop and consider

meaning, a chance to evaluate what to do with one's remaining life. In a land that is foreign and unsafe, a patient can examine why he lived the way he did and perhaps choose a future life. Richard's version of recovery was to return to what he was, an artist, as if nothing had happened. Richard wanted to get back to work badly. For him, work shut out worry to some degree. The sick always want to do more; once diagnosed, they need to live efficiently.

Recovery is not always a beautiful circle. Illness offers a less complete geometry; one does not necessarily return to the starting point. Richard had recovered from the violent exposure of surgery, but he was still ill. It scared him to think he'd be talking about his illness six months or six years after the surgery. Illness is the paradigm of being out of control. Control is a matter of making demands, and one cannot demand to get better. But back working in the studio, Richard made other demands; he was again a provocateur. He bought his first computer. He bought a scanner. He hired an assistant to load Adobe Photoshop and teach him the digital photography program. I knew he felt better. He was proud of the new dramatic scar that ran alongside his nose from tear duct to left nostril. He was eager to show it off, despite what still grew behind it.

With his strength returning, he held Cecil the cat in the crook of his arm like a long-haired football, ears flattened, tail twitching. "There's my big boy," he crooned. With his perfect memory, Richard had always been a repository of awful times. Watching him with his beloved cat, one wouldn't think that there were cold moments when cancer returned to his mind and stuck there. But he did share his postsurgical misery at opportune moments, like when he needed me to run an errand for him. He clung to the harsher bits of his illness not just as a warning system but to make the rest of us alert to his new and precarious status.

Part 3

LOSS

When Luke's surgeon first told him that the results of his pathology were fine, Luke was afraid to believe him. "But you can't be absolutely sure, can you?" His fretting question revealed his worry that he might have underestimated what was going on. His body had betrayed him, and Luke—a freewheeling, pool-playing, optimistic, always-ready-to-find-a-party kind of guy—had lost certainty. Loss is an inextractable accompaniment to and necessary constituent of illness. When the ill person visits another country, he always leaves something behind.

Not long after I last saw Luke, I gave thirty-five-year-old Maria some bad news. Maria was a competitive dancer, and I could never imagine her being in a bad mood. When she learned she had sarcoid, an inflammatory illness that could affect her brain and her heart but at the moment only involved her lungs and was provoking her dry cough, she moaned, "If I can never dance again, the joy in my life will be over." She was being melodramatic and, like Luke, asking me to guarantee that she would be all right, that if she stepped onto the soft ground after leaving my office that April morning, it wouldn't pull her into the earth. When I told her that she was in an early stage and most patients had only minor symptoms for decades, Maria, like other tourists, made it clear that she saw no point in focusing any further on illness because she was just passing through. She had been quick to jump to the worst outcome, and just as quickly she decided that her sarcoid symptoms would soon be gone. There was no need to settle down into illness; she was moving on, not staying put.

Commonly, patients experience losses more gradually than suggested by Maria's dramatic statement about her life being over if she couldn't dance. When the memoirist Jennifer Estess first developed amyotropic lateral sclerosis (ALS), a uniformly fatal disease that affects nerves and muscles, she had no overt physical loss. Yet her muscle spasms were a signal that something was seriously wrong. When she became unable to shave her legs because her hands and calves had become too jumpy with misfiring neurons, she made excuses to friends and family. She told herself that if she tried hard, she could make herself better. She was thirty-five years old, confident, and self-possessed. But when she climbed out of a taxicab and lost her balance one day, "instinct ordered me," she reported, "to leave my body and supervise from above. My spirit hovered over the scene, trying to make sense of a new phase—and what I sensed was going to be a whole new life." When she could no longer walk confidently, she lost her eagerness to go out onto the streets of the city she loved. When she couldn't pinch tweezers and a bag of muffins felt as if it weighed a hundred pounds, she lost her optimism. When she relied on others, she lost her independence. She had lost faith, and the words she used to whisper, "I think I can get out of this," seemed pointless.

The losses brought on by illness are endless. A friend of mine, newly diabetic, lost his birthday because he spent it in the hospital after his diagnosis. He lost his favorite food: key lime pie. He lost his fantasy of good health. After some years, he began to lose his vision, his job as a finish carpenter, and his sense of power and order. He developed vascular disease, and the pain in his calves limited his walking, which took away some of his independence. He stopped having erections and was ashamed around his wife. Newly self-centered, working to control his condition, he often ignored his children. He didn't have the strength to care for his elderly mother. He was meticulous in his self-care, following every rule, yet his disease progressed anyway. He lost the time spent traveling to doctors' offices, waiting, having tests, explaining his symptoms again and again. The small toe of his right foot had to be amputated, and afterward he asked friends not to visit; he eventually lost them. Illness was a continuous threat from which there was no escape, but he was never ready to give in. Inactive, he never relinquished a sense of energy and effort. Maybe he saw his glucose level as a test of character. Maybe the diabetes reinforced all the characteristics I had always known him for: self-reliance, industry, perseverance. But mostly, his illness created a madness of mundane worries. Can I get to the bathroom quickly enough? Can I keep down this meal without vomiting? Can I walk these stairs without being gripped by pain? Can I sleep through the night?

Richard had ignored his symptoms for more than a year to avoid the possibility of *any* losses. After he received his diagnosis, he ignored that too for months. Like everyone, he hoped from the beginning that his doctors were wrong: first about the rare diagnosis, then about the likely progression of symptoms, then about the risks of treatment. But he agreed to sinus surgery as the only way to know the full extent of his cancer's spread and to remove the tumor that was entangling his facial nerves. When he was in the operating room and my sister, his daughter, and I were waiting in a grim room that had the debris of a bus station, I thought of Susan Sontag's discussion of the "aristocracy of the face," its privileged status as the acknowledged home of beauty and character and physical ruin. Historically, illnesses that transform the face have aroused the deepest dread: deforming, ulcerating, emaciating conditions such as smallpox, leprosy, AIDS. Sinus surgery was a particularly gruesome form of trespassing, and Richard understood that once scarred on his face, he would forever appear to be a foreigner, a traveler in the land of the healthy, or at best an expatriate. It would be obvious that he was an alien, that he'd been sick.

"Our very notion of the person, of dignity, depends on the separation of face from body, on the possibility that the face may be exempt, or exempt itself, from what is happening to the body," Sontag wrote. I didn't understand at the time that by delaying surgery on his face, Richard believed he was putting off his own deterioration. Once he had surgery, he would no longer be able to lie to himself or anyone else. He would have to admit that he was seriously ill. In retrospect, his delay seemed reasonable, although eight years ago I pleaded with him to have the surgery done immediately so that he could go for a cure. He wouldn't talk about it. He didn't answer me or his wife or daughter for months, although each of us knew how intensely he was feeling the misfortune of illness. Richard was a man of certainties, and he didn't like the ones he was being handed. Maybe there was nothing to be said until he was ready to take care of the problem, but Richard had had no practice in talking about what it was like to be sick. He had had an emergency gall bladder removal twenty-five years before, but otherwise he'd been healthy.

After the sinus surgery, people asked him what had happened, and Richard had to tell them. He didn't like talking about personal matters, but his illness was the only saga Richard got to speak of for a while. Listeners wanted the whole tale, from the beginning straight through—everyone liked to hear what went wrong and how the wrong was righted. He was a vocal stoic and would relate his story with fervor, with optimism, in bits and lunges. But healthy people are always embarrassed by sickness, and friends who saw Richard's scar felt bad for him. "What counts more than the amount of disfigurement is that it reflects underlying, ongoing changes, the dissolution of the person," Susan Sontag wrote. The most feared diseases, Sontag noted, "are not simply fatal but transform the body into something alienating."

Leila was an eighteen-year-old, black-haired, long-legged high school soccer star with a lovely smile and a salivary gland tumor, an expanding growth just below the angle of her left jawbone. I had known Leila since she was a little girl with Pippi braids; I had coached her at soccer when she was ten years old. She favored sleeveless black T-shirts, slightly faded. She used to walk past my house with a lively pack of middle-schoolers. I envied her ease. The year before, she had gone out of her way to say hello to me in the market. She was at an age when she didn't have to give a thought to her happiness.

The surgery was to be performed by a friend of the family, and Leila's mother asked if I could see her daughter because the surgeon had requested that she have an internist's clearance a few days before the operation. Leila had always liked me, her mother said, and the family would be very grateful. It was May, and as a father of my own teenager who was ready to get out of school for the summer, I was concerned that Leila's romantic life would be foremost on her mind with neck surgery pending. I didn't see many eighteen-year-olds in my office; they were generally healthy, and having graduated from their pediatricians, they stayed away from doctors. While Leila was changing into a gown—she required a quick examination, an EKG, and blood work—her mother and I stood in the corridor just outside the room.

"Does she have a boyfriend?" I asked.

"What do I know, I'm just her mother," she replied.

I had spoken with the surgeons, and I knew that a part of Leila would be cut away, leaving a glossy divot the size of a golf ball on the left side of her neck. She would be disfigured either by disease—her tumor, though most likely benign, had grown quickly—or by the treatment. Fortunately, her scar would not reach the privileged sanctuary of her face. It would stop just under her chin, maroon and glossy, but would be difficult to cover, even with a turtleneck. It would, in the days immediately following surgery, match Kafka's description of the child's wound in "A Country Doctor": "Rose-red, in many variations of shade, dark in the hollows, lighter at the edges, softly granulated, with irregular clots of blood, open as a surface mine to the daylight."

Certain conditions are rich in symptoms—coughing, rashes, swelling. But most often, illness is inside, invisible. For many, illness is an abstraction. It is what the doctors are talking about and what is treated, but patients may not have a sense of the disease in their bodies. Lung disease, hypertension, diabetes, kidney cancer are all silent and unseen. A heart attack is not physically transforming. Thus, it is possible for patients to pretend they don't have, and never had, these conditions. But disfigurement can never be hidden. One has a new physical identity that is apparent to all. As social beings, we gain first impressions from the surfaces of bodies. Our exteriors—lipstick color, tie pattern, skirt length, the height of heels, a baseball cap worn backward or forward—determine whether people take us seriously. Oscar Wilde suggested that only superficial people are not concerned with appearances. For Leila, her scar would be like a scandal, a public weakness, something she brought on herself.

Patients feel betrayed and therefore shamed by illness, and what is shameful should be concealed. But disfigurement makes this impossible. I knew that for many years to come Leila would have to run into the dark and hide to pretend she was whole. The morning after her operation—she would go to sleep in one country and wake up in another—Leila would feel nakedly flawed. Each morning for days or weeks after, she would awaken and look at the red geology of her formerly tubular, tendon-defined neck, and see it as a compacted, dense site of despair. She would have to decide how her new appearance was going to transform her. The suffering of the disfigured is a problem of management and self-definition, of choosing how to see oneself.

I had seen dramatic, disfiguring losses, but I wasn't going to tell this to Leila. In my first years of training, I met Ray, who had been in bed so long with multiple sclerosis that his body had shrunk and his limbs had contracted. Although he was not in physical pain, his legs were permanently bent under him like wings. He couldn't tap his feet in nervousness. He couldn't put his palms flat on a table. When the nurses rolled him over to clean his back, I saw bed sores that gave off a peppery, oaty smell.

"The quadriplegic will be forever dependent on someone," novelist Andre Dubus, an amputee as a result of a car accident, wrote. "He cannot sit on a toilet, he cannot wipe himself, or shave, shower, make his bed, dress. He will use a catheter. He cannot cook. He will not feel the heat of a woman, except with his face." Dubus counted himself among "those who *cannot*," persons whose spirits are willing but whose flesh refuses. I felt embarrassed around quadriplegics, discomforted by being whole when they were not, and at first I was reluctant to talk to Ray. I went in and out of his room (he had been in the hospital so long that I thought of it as Ray's room) with the big Amnesty International sticker on the wall and did my job cursorily, ordering tests and medications. It took me almost two months (I had inherited his care from another doctor who retired) before I actually sat down beside Ray's bed to learn what he had to say. Before this initial conversation, I told colleagues that the most that could be said about Ray was that he had committed himself to *remaining*, to staying alive because it was all he knew how to do, even though his body no longer functioned. I was afraid that if I heard his story, it would be long and sad, and it would be a long and sad process trying to console him, if he accepted comfort at all.

Like all young doctors, when I witnessed something awful, I didn't want to think about it a moment longer than I had to. Illness could be horrendous—the strangulation of asthma, the drowning of heart failure, the subcutaneous gas of gangrene. If I did stop to consider these conditions, it was only as examples of pathophysiology in extremis. Any further analysis was conducted purposely without self-awareness or perspective. Days later, when I started to be absorbed by the sadness of what I'd seen, I'd secretly and guiltily begin to think of the patient as a person who had *done* something wrong instead of simply having a body that had gone wrong.

I wanted to put Ray's losses out of my mind instantly. Over the years I have come to see, however, that there is nothing doctors are allowed to put out of their minds. A doctor's job is to offer unyielding attention. Pride and a sense of competition allow one to function among patients for whom nothing works, who hardly care if doctors are around. Medical training sensitizes one to horror and permits a sureness before awful sights. Nowhere is this clearer than in dealing with disfigurement, which has helped me to grasp the true frailty of bodies, the fickleness of the compromises we all make in self-image, the deadening effects of self-loathing.

I assumed Ray was bitter after being bed-bound for six years. I supposed that he wanted other people to have bad luck too, and that he wanted others to know that what had happened to him could happen to them. We don't know what patients are thinking, however, even if we believe we do. Curiosity needs to be everything for doctors, but I had refused to be curious about Ray for months. I remembered that the first time I met him I found this passage in Job that evening: "The thing that I so greatly feared has come upon me."

The day I decided to sit down and speak with Ray, the new

spring leaf buds outside his window had turned the trees blurry. After a heavy rain the night before, the windows were streaked and visitors were few, the bad weather an obstacle. The hospital felt empty. It was early morning, and breakfast seemed far off.

It turned out that Ray had a sense of humor about his plight. "Isn't one of the guarantees of American citizenship that each of us gets one medical miracle?" he asked me. He didn't cry for himself but instead for other people. He watched CNN continuously and took in the news of people beheaded, burned alive, starved, tortured, and he was always moved. He raged against cruelty. Ray used the widest scope of human atrocities to keep his own despair in perspective. If I had compared his plight to those who were "worse off," I might have been accused of denying his suffering. He was taking on that role himself and sparing me from that awkward position. In some ways, it was a gift to all visitors; from then on, I would never fall into the silent game of weighing Ray's discomfort, unhappiness, disfigurement, loss, against any other sick patient I saw that day, that week, that season. I wondered if happiness depends on the fact of good health or rather on a view on one's health, during or after illness, that accepts rough-edged truths. I respected Ray's illness because he respected others'. He became my standard for bravery and the expression of shared grief.

After Leila's EKG, she wanted to know about other people who'd had her type of surgery. She wanted to hear how they were doing. At first, I thought she was asking about the extent of their pain, but really she was asking about prognosis. Had they all recovered? Had they all moved past surgery without further harm? Her questions were a means of taking her problem away from her body and attaching it to someone else's.

Illness exposes a disunity between the quintessentially unified mind and body. The ill person sees herself both from a distance and from within, and she moves back and forth between these vantage points. The ill body is sometimes split off from the well mind as the mind estranges itself. When we were first talking, I heard Leila say, "My *neck* needs surgery"; later she said, "*I* need surgery." This reminded me of Joanna, who would say, "My left foot hurts," more often than, "I hurt." In the language of illness at least, body and personal identity are sometimes discontinuous. The hyper-reality of illness ruptures an innate and natural unity. The body is a discrete entity, an "it," a machine, separate from thought and emotion. Emotions—fear, sadness, hope—correlate with bodily function but not exactly. "I was what things within me happened to," the novelist Paul West wrote as he waited to receive a pacemaker. His body was a secularized domain. "I felt very much outside myself, allowing them to do all this to someone else into whose private thoughts and sensation I could peer." Doctors certainly contribute to this feeling by segmenting the body into that piece that is the territory of interest and the other parts that are beyond their interest or irrelevant at the moment. But patients naturally do the same thing to themselves. It is a way to *contain* illness and to distance oneself from it.

I can tell from the shy downward cast of a patient's eyes and his soft murmurs that he trusts me. But trusts *what* exactly? That I am good at my job? That I know what he is feeling? That I would put myself in his place? That I know how to comfort him? That I will avoid telling him anything too painful? If he trusts me, I also know that patients often blame their doctors: How could this betrayal have happened to me? Why weren't you watching? How could you let it happen?

I was Leila's former coach, and she trusted that I knew how important soccer was to her. She wanted to know how soon she would be able to play again. Just moments before, I had wondered if Leila was still interested in soccer; she had discovered the lump on her neck after a soccer field collision, and in my experience the ill sometimes dwell on the occasion or site of the discovery of bodily betrayal—the place where life changed—and won't return to it. I was relieved by Leila asking about a return to the field. Like most doctors, I accepted the mysterious importance of a patient's positive attitude. "How could you ask about soccer at a time like this?" her mother interrupted. I thought: *There has never been a time like this for Leila.* This is exactly the time for Leila to express hope, to wonder how things will change when she returns to the soccer field.

Illness rearranges. Everything is undone, and the new world is odd. A new diagnosis may leave a young person undirected and bereft, but sometimes an illness reassembles a person and redirects him; he has a cause. The young celebrity with Parkinson's disease suddenly represents all persons with Parkinson's.

He becomes a medical research fund-raiser with his own foundation. He puts his intelligence to work. He learns all there is to know about his disease. He does interviews with a lighthearted didacticism.

I was sure that Leila's mother imagined that after her daughter's diagnosis there was no more life. The linkage of *any* diagnosis to death is virtually irresistible, but I had already reassured them that outside of the small operative risk, Leila's tumor, despite its unfortunate location, was not fatal. Without the fear of death, there was still worry and heartbreak, in a mother's view. After illness, "there is something else, something stumbling and unliveable," as Lorrie Moore has written. Leila's mother had a sense of her daughter's imminent departure into the land of sickness. It would be her daughter's first trip to that foreign place where, as Harold Brodkey wrote, "the needle has replaced the kiss." Leila's mother had an experienced traveler's reaction to the preparation for surgery: anxiety with a whiff of disaster. When she took me aside to discuss Leila "pulling through," I had an image of her daughter riding a jeep stuck in deep mud.

Medicine is filled with rituals. A tube placed and removed. An incision sewn. Doctors leave marks on the body. These marks change the value of the body. Scarification is tribal and, in certain parts of the world, increases one's value. It is a rite of passage into adulthood. Here too: surgery can be a sign of life experience, evidence of maturity and experience. The sternal suture line of open-heart surgery, the flat plain of a breast taken away. Rituals remind us that the stories we tell of illness are not moral fables, and they are not sleight of hand. The body has a history. Disfigurement is more than the transitory inconvenience of surgery or chemotherapy or a radiological scan or an intravenous line insertion: it is permanent. Leila would literally be "marked" forever.

Unlike Luke, Leila was unafraid of surgery and wasn't worried that she was going to die. She did not expect to be deformed. She was doing what she was told to do, and she expected to wake from surgery with her life uninterrupted. But the disfigurement of surgery, like that following an accident, can spiritually age patients. "The change, of the spirit's immersion in horror, may cause a state of detachment from people whose lives are normal and who receive mortality's potion, drop by tiny drop, not in a torrent," wrote Andre Dubus. It would have been natural for the torrent to sweep Leila to a place unmarked on any eighteen-year-old's map. I was doubly concerned because Leila had been raised never to ask, but to accept what was offered and say thanks. The hospital is a country of lovely smiles, and Leila's, even on the morning of surgery, was particularly light, unrepentant, amazing, and somehow this made me confident for a full and quick recovery.

In the first days after surgery, Leila had to deal with the pain. Hers was quiet, insistent, and ignored at peril. She never screamed to the nurses for help, as the others on her ward did.

But she didn't like the blood that seeped into the bandages on her neck; blood was worse than pain—it was serious. Leila had a personal philosophy of no tears. She also had discreetness and an idea of control. Some patients stay quiet to preserve their strength; at a certain point, recovery is a matter of nothing more than acquiescence. But Leila also wanted to be liked; she wanted to be the perfect patient, just as she was the consummate midfielder.

She was competitive. She didn't want to lose. She was a coachable athlete and a conscientious hospital patient. She didn't resist what she was told to do—not to turn her neck quickly, not to wet the bandages during her sponge bath. Small-framed, she was polite and strong-eyed. There was nothing hazy or slippery in her answers despite her pain medications. Successful athletes are stable; their reactions are not extreme. Leila resisted extreme unhappiness during the two days she was in the hospital. She accepted the changes of her dressings, the smell of warm pus and blood and new gauze. She didn't want to lose her strength. A few friends came to visit. As is often the case, some whom Leila considered "close" didn't show up. There are always those who wonder about the patient from afar but prefer not to get too close. I visited, although she was a surgical patient and my input was not required. Her mother brought in her favorite foods: pasta with red peppers, eggplant grinders, and Gatorade, and Leila ate hungrily after the first night.

On the field, she had always been loud and confident, but in the hospital, a transformation occurred. When I visited her, not only had her physical self changed, but Leila had gone from tough and scrappy to compliant and cool. At first, I thought this was due to her medications. Then I considered the possibility that Leila was being sensitive to her surgeon's cues. Perhaps he had indicated that she should not get upset, that accommodating to her relatively simple operation was the brave thing to do. I thought that this transformation might also pertain to Leila's age; she was deciding who she would be, trying out different personas and philosophies and attitudes toward the world. What the surgeons had done to her neck was not a sign of brutality; it was awful, but a necessity, she told the nurses. Then I recalled Lucy Grealy's behavior toward her doctors: if there were to be other operations in the future, she wanted the doctors to think of her as the best patient, the easiest and most grateful, and therefore the one who deserved the best care. Leila seemed to understand this intuitively. Also, she was now scared in a different way than she'd been before. The tumor was cut out, but would it come back?

When patients are in the hospital, they read junky magazines because they have given up on thought; they refuse deodorant because they have given up on vanity; they watch television rather than speak at length with visitors because all they want is selfish ease; they eat hard candy, the more garish in color the better, because it is so clinically incorrect and unhealthy that it represents a lack of inhibition and is a purposeful affront to illness. Leila never ate candy, she told me. More evidence of her discipline. Even when shaking my hand at the end of my visit, she had the bodily self-respect and concentration of an athlete. She was at home in her body; she always depended on that freedom of movement. But Leila could no longer touch her chin to her chest. A simple nod hurt. Hunching her shoulders to her ears made her wince. She thought this was temporary, but in the days to come her scar would thicken and firm up, her skin would tighten, and flexibility would diminish. The quick snap needed to head a soccer ball was in jeopardy.

The body of any athlete knows instantly that something awful has been done to it. Leila could put up with nausea, soreness, and constipation. But I imagined she resented that the secret place of kisses and nuzzling had been trespassed.

Leila was interested in beauty and glamour. Scattered across her bed the day after surgery were copies of *Elle* and *In Style* and *Marie Claire*. There was also a magazine called *Shape*. Whoever had delivered this reading material probably had no sense of irony and maybe even thought the magazines would be helpful. After all, Leila didn't look diseased in her bathrobe and bandages; she was just a tired young woman disguised in scrubs whose hair, black and always a little wet, as if she had just stepped from a shower, was already growing long so that it would cover her neck. Leaving a copy of *Shape* was like saying, "Get well soon," but it made no sense; "soon" and "shape" were far away. Whoever sent *Shape* had not been poisonous, but that person had also not examined Leila as I had. Her neck, like the road where a land mine had detonated, was confusing to look at, a site of damage where inside and outside blurred together. It was as if some enemy had wanted to see Leila's throat exposed, and the surgeons obliged by opening her neck and letting it emerge. Her wound affected me powerfully that morning.

One challenge of doctoring is to thwart the patient's urge to collapse. In the days after surgery, I was sure that Leila wasn't anywhere near collapse. For every patient, waiting becomes an activity, an exercise in abeyance. Patienthood is confinement. Bored, patients strive to counteract the enforced affliction of backless gowns, the restriction of movement. There are formless hours to be gotten through, useless flowers, obscure relatives who visit briefly, school being missed. For Leila, idleness took on its own sharply focused purpose, as it often does for teenagers. She read her magazines and grew quieter, even as she regained her energy. When I asked her how she felt, I had the sense that she felt accused of something terrible and didn't know how to reply. So I didn't ask again. What I had gained after medical school, with its expected list of questions to get through with every patient, was a feel for when to stop talking. Over the years, what had impressed me was how much people *hate* being sick; there is real outrage and anger that often transforms after treatment into a sullen, silent discontent. Leila's silence was like a gloved fist. The natural state of being a patient is bitter disappointment combined with the loneliness of being unable to talk about what one is thinking. Leila was a teenager—approval or disapproval defined her. She was monosyllabic and wanted to go home.

As the memoirist Jennifer Estess lost her ability to sit up or dial a phone, as she lost the muscle tone that allowed her to laugh or speak or breathe, she told her sisters that being sick was like "drifting farther and farther away on a raft." I didn't want Leila to feel abandoned, adrift in another country, unreachable. I told myself that Leila didn't have anything as awful as ALS; she had only lost the perfect contour of her neck. I had certain canned speeches for patients, and I heard myself using them with Leila. I offered them slowly, and with purpose, but they were pep talk—nonsense—nonetheless: "You need to just make up your mind to begin the necessary..."; "You have to get used to the idea of..."; "Here's what I expect of you...." Word packages, I called them. But when I avoided this patter, I was embarrassed by my inability to lay out for Leila the upcoming struggles that other disfigured patients had shown me. The fear that the scar would come undone and they would split open, turn inside out. The worry that they smelled peculiar, that they carried a faint odor of decay. When I tried to describe what Leila might experience, I knew it would only imperfectly reflect what would occur.

When Leila left the hospital, she was supposed to go home and take it easy. Her first visit to my office was two weeks later. Her mother again called to request this appointment, reporting that the surgeon's postoperative visit a week earlier was "too short." Leila walked in that day as if she were cringing, self-protective, labored, determined. Unnatural.

To be a teenager is to be vain. All teenagers worry over the opinions of others. When I sat Leila on the examining table across from the three-foot-tall vertical mirror, she sat silently without moving. She kept her eyes down. She was not sure what I would do. I had suggested this two-week visit to my office to give Leila at least half an hour when she didn't have to be brave. I had taken to heart Arthur Frank's observation: "I have never heard an ill person praised for how well she expressed fear or grief or was openly sad."

I took in the shape of her head, the blue collar of her blouse, the defeated angle of her chin, her flushed face. I leaned toward her, beginning the intrusion of the physical examination. When my hands came near the scar, she seized up. I knew she felt exposed and disturbed, so I moved slowly. She wanted to resist, not let me touch her, but when she flinched, she saw it as weakness and a moment of self-pity.

"I really don't like this," she said. She was almost inaudible. When she looked up, Leila was red-eyed, close to boiling with tears. She started talking. "Why couldn't it have been on my leg or my stomach, someplace under my clothes? I'm misshapen, deformed. I dream about having my neck back, and then I have these really weird dreams about masks and veils. After school, I go from drugstore to drugstore, searching for makeup to cover myself, to make my neck secret." I understood that Leila wanted to be glamorous. Leila wanted to feel like she could be in *Elle* or *Shape*.

I remembered that during those first days in her hospital bed she avoided mirrors. Like Perseus, who saved himself by not staring at Medusa, perhaps there was a similar strength and intelligence in Leila refusing to look directly at her illness. To gaze at the horrendous face of illness, at the hard outline of deformity, can turn a patient to stone.

Only after she left the hospital did Leila learn just how many reflective surfaces there are in the world. She spotted herself in the silvery sides of office buildings and spoons, in the side windows of cars, in the security camera monitors in stores. The sunglasses of friends contained her image. Leila had come to see herself as a different person from the one she knew before illness. And she didn't like what she saw. She was petrified, scared stiff. Illness is weight—it is heaviness and constriction.

Once upon a time, Leila had owned and possessed her body, but she had turned it over to doctors, and they had altered it. In my mirror, she gave herself a startled and fierce look. She told me she thought of her neck as strange, unfamiliar, not hers. There had been two weeks of absolute nonrecognition. Those rare times when she touched it, the scar was mute and insensate, not live tissue. It was not a neck like you would draw on a piece of paper, perfectly designed, clean-lined. It was unnatural, unexpected, not resembling what should be there. She had a constant sense of mismatch; there was an incongruity between what she was and what she saw, between what she *thought* she was and what she found in her reflection. She felt deceived and stupid. "More than the ugliness I felt," Lucy Grealy wrote, "I was suddenly appalled at the notion that I'd been walking around unaware of something that was apparent to everyone else. A profound sense of shame consumed me."

For Leila, getting dressed had become an exercise in concealment. Scarves, turtlenecks, soft cotton, cashmere. In the mirror in my office that day, Leila was forced to look at her startling new topography, but I was there as a reminder that this was superior to the salivary gland tumor that once swelled her jaw. The minutes with me were torture. I stared at her throat and squinted and showed concern and took notes. Without a visit to a doctor, without mirrors, patients can harbor fantasies of nonseriousness. With a mirror, there is no escape.

When did she start to despise her body? When she realized people were staring. When she realized people were not looking into her eyes but at the destruction under her chin. Some were looking quickly, then taking second looks. Others were leaning forward to stare. The stares felt like accusations. They were not discreet. Sometimes when a patient is ill, inhibitions fall away, and overcome by needs, he is unaware of the intrusive looks of others. But disfigurement is about being *fully* aware of the vanquishing looks. "They're shocked at the sight of me," she said. "The last time they looked at me, I was normal." Strangers pointed, laughed, passed judgment. She felt powerless and like a failure. Leila had always been pleased to be in the spotlight, to take the penalty kick. But now she yearned to turn people's attention *away*. Leila had been an extrovert, energized by people around her. But now she had lost her sociability. She found people dreary. She preferred to be alone, and the loneliness of her self-imposed isolation only validated the strangers who studied her neck when she ventured out.

Leila had stopped being interested in her body. Her body was like a cat she didn't want to hold or get to know, and she acted cruelly toward it. She had adopted an attitude of paranoia, where every whisper was a comment about the way she looked, every laugh a joke at her expense. Being observed was like being put to a test. Leila had to concentrate to make her tormentors disappear, had to fill her ears with noise so she wouldn't hear the comments.

The physically odd never fit in unless they imagine the world is populated with other odd people. When ten-year-old Lucy Grealy left the hospital with half her jaw missing, her face a triangular shape, she believed that "the only reason people stared at me was because my hair was still growing in." But Grealy had a poet's eccentric, observant eye; Leila had never seen the world as odd. She bit her lip; she was a small child in constant fear of punishment. She felt like a loser.

"How's school?" I inquired, innocently burying this difficult question between asking her to raise her arms above her head and asking her to touch her chin to her chest. I saw Leila's shoulders tense; I could tell the very subject made her feel cornered. Out in the world, stigma is never minimized. For Leila, every exhibition was a risk. It wasn't as if she didn't understand why people were staring. It wasn't as if she didn't understand the urge to run away when one passes a person who is asymmetrical, who looks as if a piece of her has rotted off.

"School's not so great," Leila answered. "Kids stare so that they can feel better about themselves."

I wanted to say: all you have to do is walk away. I wanted to say: adolescents experience their bodies changing, and your peers are seeing in you *their* fears of these changes gone wrong. I wanted to say: they can see what's happened to you *on* the surface, but they're wondering what's happening *under* the surface, where they believe another reality is hidden. If the "deep truth" is inside, a scar is only part of the story. I wanted to say: you don't have to be perfect. But it was too late to warn her.

John Merrick, the "Elephant Man," fantasized about living in a lighthouse—his conception of seclusion—and also about life among the blind. "There was a half-formed idea in his mind that he might win the affection of a woman if only she were without the eyes to see," reported his doctor. Merrick was a small man with an enormous, mis-shaped head. Projecting from his brow was a huge bony mass that occluded one eye, while at the rear of his skull hung a bag of spongy, elephantine skin. A pink osseous stump turned his upper lip inside out, and his nose was recognizable as a nose only by its position between eye and lip. The circumference of his head was that of a man's waist. Exhibited as a circus monstrosity and object of loathing, the public exposure of Merrick and his deformities was said to transgress the limits of 1880s London decency, and his show was closed. Walking the street, he wore a peaked black cap with a gray flannel curtain that hung in front of his face.

The worst irony was that he had hip disease that denied him all means of escape from tormentors. Unlike Leila (who knew his story and also wanted to be free from the ring of curious eyes and the whispers of fright and aversion), the Elephant Man could never run away.

Leila told me that all the unwanted attention she received reminded her of girlhood teasings from the playground—years not far behind, when differences marked you as a freak, a monster from outer space, or a witch deserving of elimination. "An ill child withdraws when he senses that people do not like what he represents," wrote Arthur Frank. "To other children his presence brings a fear of something they understand only enough to worry that it will happen to them." We all have some buried fear of contamination when we are with the sick. Mortality is catching, and being near an ill person makes us susceptible. A scar, the novelist Margaret Atwood wrote, is "where death kissed her lightly, a preliminary kiss." We can see the mark.

Because illness takes place on or in the body—which we expect to function perfectly—all illness feels like personal failure. At school, Leila had points taken away for the wrong answer; her senior year felt like going from one exam to another. Now her body was a test: part of her had been taken away. Her body was wrong. It failed when before it had passed.

Telling herself she shouldn't be ashamed gets a patient nowhere. It is inane for a doctor to tell his eighteen-year-old patient that shame should be reserved for mistaken actions and poor choices. Humiliation is the loss of dignity. Leila felt that her situation was not tragic, but dishonorable. A short time before, she had assumed she was unique; now she had arrived in the land of sickness with vague feelings of fear like anyone else. At the mercy of others, Leila had grown contemptuous of herself. She wanted to disown a part of herself. She emptied herself of feeling altogether except for repulsion, disgust, and sadness.

Disfigurement has a particular shame because it is hard not to think of it as an emblem of decay, as having a moralistic meaning. This emblem doesn't appear for something you did, but rather for what you were, or had. Leila was a young woman who was damaged. She wanted her friends to pretend nothing was different. She made an effort to help them pretend by wearing her hair long and her collars up. "I want to go to school and see people, but I don't want to be seen," Leila said. All disfigured persons withdraw in part because they believe others would be happier if they did not have to look. Withdrawal is a deal: I will disappear to keep from causing you pain as long as you won't come looking for me when I'm gone. Leila started to pull back and stayed home more than ever before.

Leila stared out my tall windows at the ceaseless threat of sudden spring rain. Weather provided drama from the fifth floor. The clouds felt intimately present. Soon rain splashed against the glass. "I wish people would just say, 'You look awful,' which I do," Leila told me. "But I guess no one's going to say that." Then she became more emphatic. "No, I wish they would scream, 'My God, what happened to you?'"

There is almost a taboo against asking this question of the disfigured. We appreciate when children do the dirty work. We know children are asking not out of loathing insinuation but with gaping, innocent curiosity. A conspiracy of silence exists around all illness, and this is particularly pointed when disfigurement has made the patient's losses so obvious. The disfigured person cannot be lied to, so silence is the rule because no adequate comfort can be offered. Perhaps we don't ask because we don't want to hear the awful details. What we learn will be so awful that it may change our interaction with this person forever. Perhaps we don't ask what happened because we don't want to hurt the feelings of our disfigured companion and make her feel vulnerable. Perhaps we are afraid we will be overcome with pity. Perhaps we are afraid that the question *What happened?* will be interpreted as *How did you get to be so ugly?* Even if we are simply curious, perhaps we remain silent because we are never far from the words "abominable," "grisly," "repugnant," "horrid," "grotesque," "obscene." We start thinking, *How unfair,* and end thinking: *Consumed, shriveled, invaded, reduced, mutilated.* Perhaps we question our own curiosity: What is wrong with me that I want to know? Why do I want to know? So I can save myself from it?

Although Leila reported that her scar felt like it was lodged *in* her, it actually looked quite superficial. Uneven, discolored, it spread along her throat and under her ear, but gripped her like a choke hold. Leila didn't like touching her neck. "It's a mess," she said, the heat of surrender in her face. What she meant was either, *This is emotionally messy,* or, *I'm a mess.* She wanted her life tidied up. She wanted her problem scrubbed away. Part of her had been taken away, and I imagined that when she touched her neck, she had a sense of spatial loss (*Where is my full and entire neck?*). This emptiness must have caused a specialized fear, a disorientation. The body she knew was no longer the one she had. Disfigurement, like pain or terror, is isolating, and to counteract this sense of separation the patient thinks: *This isn't bad, isn't really bad, isn't real.* With every touch of her neck, Leila discovered just how real it was.

For the viewer, disfigurement carries a disagreeable, painful message that one's troubles are never really over. It makes a lie of the maxim "All wounds heal." We are *not* clear of the events that have marked us; injuries and transgressions stay with us. We recuperate, but we never fully recoup our losses. We return home after surgery, but we never get back to where we once were, because now we are different. Leila understood that she hadn't left illness behind in the hospital; she had brought it home, and it lived on with her. Disfigurement tells every viewer that past traumas are stubborn, enduring, reshaping. This is the power of disfigurement: proving that history is ubiquitous, it sculpts us.

All of us have a certain image of who we are. We brush our teeth and hair and every morning say, *I look like myself.* Few patients ever fully accept that they have been physically changed by illness. Once illness has passed, they underestimate or simply distort any changes they see when they study the face and body in the mirror, thinking, *The suture line is barely noticeable,* or, *My hair has started to grow back*. They pretend they are perfect. At night, they do not dream about missing pieces of themselves but about being complete. Their faces and bodies are fixed emblems, and a considerable metamorphosis is rarely enough to change this view of themselves.

When Leila stopped crying, she held very still before the mirror and tried to calculate her new worth. She took in her face, her bones, her eyes. But whenever she looked at her scar, she was studying how she *looked*. She tried to identify the essential *Leila*. The disfigured patient travels toward the interior, into her own depths, away from the telling surface; beneath her scar, Leila saw her faulty character, her loneliness, her future, the evidence that no one could protect her. She was dealing with disappointment, with the reality that she would never be whole, that she would never look like other people or look like she wanted to. As often as she checked that day, there were times when she was surprised by the sight of herself. She realized that she hadn't had a strong sense of what she looked like before surgery.

I wanted to tell her that her neck looked fine: the new skin, thin and pink and delicate, had healed beautifully, and in the curve of a shadow, the smooth indentation was barely noticeable, hidden amid hair and tendons. I wanted to cleanse her of the ridiculous curse of embarrassment. But for the disfigured, there is no promise of solution or victory. I wanted to take Leila to meet Ray—he was a master at negating despair and maybe she could learn from him. But for Leila, new to illness, despair was not hierarchical, as it was to Ray. No matter what the fate of others, it was acceptable for Leila to suffer. It was right, I realized.

By the way she jumped off the table, I was reminded that Leila was still very young. I wanted to give her a lollipop wrapped in cellophane, to wink at her and tell her a little joke. But she was not a little girl; she used lipstick and an eyebrow pencil. She was eighteen years old and angry. She was damaged and toughened. Even if she wanted to be private, her scar made her public and unforgettable.

I retrieved Leila's mother from the waiting area. She had on heavy silver earrings that carried amber beads and a matching necklace that lay over a loose-fitting tan blouse. She had an inexhaustible tenderness that made me think about how terrible and lonely it is for patients who lack a caregiver. Leila's mother chose to stay near, to take orders, and to not leave despite a teenager's typical scorn. She refused to be set free by Leila, who barely tolerated her brightness. She had offered her life to Leila for eighteen years. Caregivers joke through horror and unease; they shop for and cook for and feed and undress the sick one; they answer the phone; they lie down with and sing to the patient. Forced or not, the close care of a patient has what Brodkey called a "courtship intensity."

According to her mother, Leila showed few outward signs that she was bothered by her new status. She had not become cruel to others, or arrogant, or callous. She had not taken on any of these ruinous defense mechanisms. But in the privacy of my office, with her mother outside in the waiting area, Leila had admitted, "I won't watch soccer anymore." She avoided the fields she'd played on and all the cable channels televising games from Europe. She hated the carelessness and exuberance of her favorite players. She couldn't bear it; they seemed omnipotent. She was embarrassed to admit that she was flooded with spite.

The ill are jealous. When Oliver Sacks, recuperating from surgery, was filled with virulent envy of a neighbor's good health, he said to himself, "This is not me—not the real me—but my sickness which is speaking." The wellness of others produces a soreness, an irritation. While the ill person is in retreat, the healthy enter the world, tumble, dance, drive trucks, and lead a familiar kind of life. The tempting world and its panorama of possibilities feel just out of reach. Patients want to be good sports. They don't *want* to idealize others' experiences. But they do, and to counter this the ill can become judgmental. Believing that they see things more clearly than before and that they have the time to watch closely, they feel defiant—even as they also know that, in their nakedness, in their exposure, in their ongoing need to explain their condition to doctors, they too are being judged.

Leila broke down in my office because her body had broken down, because she was ashamed of herself for having and confessing her poisonous thoughts, because she knew she would soon be allowed to play soccer again but she had lost her nerve. If she went back to her team, she knew she would be apologetic—for missing games, for looking different, for having been ill. She would be an object of curiosity at best, a charity case at worst. Maybe she'd have to play better than before, or maybe people would expect more. She didn't want to apologize; illness had made her shier than she already was.

She couldn't bear her feelings, her incapacity and inadequacy. As hard as she tried to staunch her tears, to calm herself, her sobs came at a steady rate. Doctors give patients permission: to relax, to eat a sack of chocolate, to sleep, to study the Bible, to want something for nothing. Or to sob over the losses. To be overcome by melancholy and anger because it is all too much. To go a little crazy, since there wasn't time to do that during illness. But doctors also give patients permission to forgive themselves, to say to themselves: the body is disloyal but not fundamentally unlovable. Crying, Leila was admitting that she had changed and was enduring yet another loss. But she was also acknowledging that illness was not a difficulty of her own making. "I guess I'm an adult now," Leila said, offering an eighteen-year-old's view of growing up and its necessary disappointments. She wiped her face, blinked, and, like the athlete she was, readied herself for another battle.

When one is ill, it is mysteriously almost impossible to re-create the sensation of wellness. Like prehistoric life, it is something one can imagine, but it is still extinct. Health has passed out of existence; it has died.

I remember the sound of Richard hitting the floor eight years to the month after his surgery. It had been a nearly perfect Sunday morning. There was coffee and air and bright light, and my brother-in-law had given me a fond rub on the top of my head as he passed by to get himself an extra spoonful of sugar, not even castigating me for not having given him enough. When he hit the dark oak kitchen floor, the sound was dense, sandbag-resonant, with no echo. He had collapsed unexpectedly. His legs, thinned by the loss of twenty pounds (diet or illness?), just gave out. His feet lost their grip; he lost his grip.

There is a precise moment when I awaken to a patient. I forgive everything I might have disliked in that person up to this moment and see him before me as innocent. None of us had expected Richard's loss of balance and coordination, and seeing my invulnerable brother-in-law on the floor, I suddenly believed what I had not believed before. When I knelt beside him, the cancer was no longer what I was afraid of; it was the future, and I prayed. Despite his surgery, despite radiation treatment that had rotted his teeth and evaporated his saliva and made swallowing difficult, despite watching him grow skinny, it was only then that I admitted how dangerous his condition was. Rather than seeing him through playful, hopeful reminiscence, I saw him as he was at that moment: sunken, plaintive, fallen.

One moment, dazzled by the living room sunshine, I was a respectable young doctor having his weekly visit with his family. The next, I was older, another stooping relative, bent to help but wanting to retreat, reluctant when his eyes met mine. I hoped he couldn't detect how uncertain I was about his ability to face the tasks before him. His fall was not due to simple clumsiness; it was the beginning of the end.

Some imagine that the separation between health and illness is subtle, that trying to pin down the moment when a person thinks of himself as ill is like trying to observe that instant of falling asleep. The switch from the unconsciousness of health to the consciousness of sickness may be invisible, but it is rarely subtle. Illness is sometimes an abrupt transformation, but other times the patient crosses over from health into illness simply by giving up his resistance to the idea that he isn't sick.

When illness arrives, patients look surprised, as if suddenly they've been accused of something terrible and they have to think of a reply. For the first time since his diagnosis, my brother-in-law felt cursed. "Goddamn it," he screamed. Collapsed, lying on his side, he cursed back. There was fear in his expression, but also a blazing amazement. I knew that this spill made Richard feel that his illness was an insult. The coffee had wet his blue shirt and splattered a puddle near his chest, like muddy blood. The mug had skittered away ten feet toward the counter, unbroken. I asked if he could stand, and he sensed my burden and interpreted it as disappointment with him; this shame made him curse at me for taking so long to help him up. He was on the floor, I was above him: it was a cruel arrangement.

When I saw Richard on the floor, part of my mind was already working on a purely medical plane. This surprised me more at the time than it does in retrospect because I didn't *want* to be his doctor and I had avoided most medical conversations with him. But suddenly I was thinking about the tests I would recommend and what his physicians could find that might be reversible. Thinking medically made me feel useful, while being aimless, with Richard on the floor, felt horrible.

Being a doctor and being a patient have this in common: we mostly worry to ourselves. Despite all good intentions, speaking about these worries, being completely candid, is risky because it makes one too vulnerable to gloom. With Richard, I couldn't be honest with myself, and I couldn't deceive myself either. It was terrible to see him down, but I couldn't admit that he was dying. I felt like a fool.

Part of me was thinking as his brother-in-law. I felt sorry for him. I felt for him. I steadied myself. Thoughts again collided in my brain. Should I have acted as if nothing was wrong other than a slip, a stumble? Should I have looked him in the eye? Should I have let him know what I was thinking? Should I have told him that I knew he never wanted me to see him like this? Thinking as a doctor, I realized that I had never been fascinated by the rareness of Richard's cancer type, as some of my colleagues might have been. Rather, I had been disappointed that there were no clinical studies that might have guided his treatment and prognosis or at least have allowed me to research a few facts and be helpful to him. On the floor, he was already lost to me. As Richard Ford wrote in his essay "My Mother, in Memory," "Death starts a long time before it ever ends." Grief is a stupor, a thick blanket dropped over the head. Conversation at such times can only be a gesture.

Somewhere between acting like a doctor and feeling like a relative, I thought: *If there were ever a chance that his cancer would recede on its own, it's gone now.* Richard would still sit in his studio before dawn and be able to forget about the anxiety he was causing my sister, but he would never again completely banish the thought of cancer. He and cancer were a pair from then on, irreconcilable and inseparable.

"My mind would seem to leak out, rise above my trunk and limbs and gaze down at them from a helpless nearness. When can I live again in my body? And where am I now?" Reynolds Price wrote as his astrocytoma, an eel of a tumor hidden in his spinal cord, paralyzed his legs and put an end to a carefree life. "I felt like a spirit haunting the air above his old skin that had suddenly, and for no announced reason, evicted me and barred any return."

My sister, having come when she heard Richard fall, stood in the doorway with pure apprehension and watched as he tried to lift himself. She was mindful of how frequent and severe

Richard's symptoms had become recently and had told me only a week before about his nosebleeds, his nausea—problems beyond the weight loss, which I hadn't quite accepted either. When she reported these symptoms to me, I'd thought how neither of them ever seemed as alarmed as perhaps they should have been. But I hadn't said anything—out of loyalty perhaps, or to sidestep the implications, to avoid the worry and grief that such a comment could have produced.

He cursed, and cursed us more as we moved to help. Aggression is the underbelly of fear; the ill person desires improvement and fears the lack of it. Dependency makes relations with others fragile during illness. Family, friends, and particularly doctors are never helping enough, never able to guarantee improvement. The less likely the sick person is to improve, the more vulnerable he is. And the more vulnerable he is, the more precarious is his ability to control himself. Aggression also derives from humiliation. On the floor, Richard viciously shouted at me again for not helping him up fast enough. In self-disgust, he was unable to stop screaming. I saw in Richard the triad that would constitute this new phase of his illness: desire (to improve), dependency, anger. The cats, Alice and Cecil, appeared from another room after hearing the commotion and came over to rub against him. The mean creases at the corners of his eyes softened.

When, at the age of eight, I first met Richard, he would call me over to feel his muscles. He had been a football lineman in high school and was proud of his strength. He would pump his bicep, and I would touch it with my forefinger. His best pleasure was in being strong. Now, years later, his muscles had melted away. Before the fall, Richard and I never spoke directly of the inevitable decline, but wearing an undershirt, wasted, he had held up his arms to show me what he'd lost. He wanted me to touch his bicep again. When he was looking at me, he was seeing the previous years of his life. In the last six months, he had lost fat and meat from his face, his thighs, his ass. He was ashamed of his skinny arms and legs. Only after my sister and I lifted him to his feet by his elbows and sat him on the couch in the living room—where he had spent his days after surgery eight years before—was he ready to speak of how bad he looked. "If we arm-wrestled now, I *still* think I could take you," he said to me, mock-aggressively, and smiled. He asked me if his doctors had ever warned him that he would lose weight, but he didn't let me answer. Maybe they had, maybe they hadn't. It didn't matter anymore. Whatever the process was, it had already worked its way through him. He understood that something was wrong, and he understood what it was. On the floor, he had been a hunted, wounded animal: fallen, exhausted, waiting for the fatal shot.

I opened the window and breathed in. An icy wind whipped my face, and I felt restless. I was aware of feeling helpless. Would I be able to help my sister and my niece, even if I couldn't help him? As he sat on the couch, I couldn't decide whether his expression was one of dignity or disgrace, or both. I knew he couldn't tolerate the pitiful state he was in. He had a strong respect for death—what Brodkey called "the dance of particles and inaudibility"—and I wondered if he was pondering the inevitability of it all. But there was probably more in his mind than that.

We called his surgeon, who said Richard needed to come into the emergency room for a CT scan of his head. But Richard had no interest in going to the hospital just then. He had no interest in turning himself over to the control of a highly structured, dreary, but insistent institution that had given him eight good years but was unlikely to offer more. Many patients are comforted by the neat, splendid modernity of the hospital routine. They are assured that they are receiving state-of-the-art care and benefiting from advanced technologies. Although Richard liked his surgeon, a trip to the hospital would be a somber and exhausting journey, and if he could stand up and get back into a chair, he saw no reason to have an unplanned emergency visit. "He's just a surgeon. What does he know about falls? He won't be interested. I have nothing left for him to cut out."

When Richard finally made it to the doctor's office the next day, he was alternately indignant, gruff, and respectful, a persecuted martyr. His strong body had always been a bulwark against the cancer in his sinus, but now it had become insufficient. He couldn't even keep himself upright. He saw himself as under attack. But how to defend himself? What was required? He heard the first knocking of death.

His falling produced a new subject to talk about with his

wife and daughter, sometimes to the exclusion of other topics. I became involved in his negotiations with my sister over food, vitamins, and supplements. We discussed the high-protein shakes Richard had started drinking to gain weight, and he was amused that now my sister supported his intake of ice cream, if not the onion rings from Johnnie's that I still slipped into his studio. He had little appetite in general, though, and he got me to explain to my sister that he should only eat what he wanted to eat (without, of course, saying, "What difference does it make now?"), which had been his position for fifty-eight years. But my involvement was unnecessary; no one ever made Richard do what he didn't want to do.

His falls continued and were not preceded by dizziness or unsteadiness. They just happened, implacably, unfathomably. Unfortunately, they often occurred on the hardwood floors of his kitchen and television room. As they recurred, he became loud about his illness. Flannery O'Connor, in a letter to a friend with epilepsy, wrote, "I think anyone deserves a lot of credit who has it and doesn't go around telling everybody." She valued serenity, even stoicism, and certainly privacy. Joanna became angry at herself first because she was at the mercy of pain. Richard, on the other hand, with his body betraying him and his dignity in decline, turned on us, fuming at me, at his daughter, at his wife, whoever had to help him stand again. "I'm not using a cane, so don't ask me again!" We were clear targets because, as we kneeled to get our arms under his, we had power over him. In capsizing, he had been sent a chilling message: this body has a life of its own. You will follow now rather than lead. Safe passage is no longer guaranteed. I thought of the expression "falling ill." On the floor, Richard's perspective had been altered, literally.

In some way, Richard must have felt that we had *let* him fall. When the sick stop thinking of themselves, they must sometimes think how fraudulent the world is. How can those people I love proceed with their customary lives? How can they have composure, control? How could they let this happen to me? The sick see the well as smug with their plans, agendas, and ambitions. The healthy are not like-minded or even like-appareled in their peculiarly normal clothes. They are different: hair-dyed, jewelry-wearing, superficial, in the moment.

Illness carries with it "the losses to come." Although illness was taking over his body, making his mind a spectator, Richard did not accept this. After a few falls, his acceptance was fitful, but he remained doubtful about the permanence of his new state. After a few more times down and his doctor's suggestion of a walker, Richard became more resigned to his illness, with a steady, understated gloom. Proudly monomaniacal, a selfish and constant worker, he now had to depend on others. As he struggled to stand, his eyes contained the silent, doubtful plea: *I can't help myself. Can you get me out of here?*

When one's body can no longer be trusted, when self-knowledge slips and one cannot understand what happened or cannot predict what's next, other betrayals are likely to follow. The ill person awaits the next infidelity. In Richard's view, *we* were betraying him. Why didn't his daughter move from Vermont to be nearer? Why didn't his wife stop working completely and stay home with him all day? Why didn't I visit more? He wanted us around. He wanted whatever we had. He wanted my sensibility and honesty, my secret stores of goodwill and good luck. Richard began to hold grudges; I was the target of his grievances if he sensed I was being overly sentimental or had ill-conceived hopes. He made me recognize that when patients refuse to acknowledge my kindness, I only become kinder.

Serious illness includes the sick person's sensation that he will be stumbling around until death. Our view of our own bodies in decline is marked by an astonishing and definite shift in self-knowledge. When we are well, we may live close to illness, but we never bother to imagine it. Illness is like visiting a relative whom we've known all our lives, who has always been quiet and reserved, but then one day is nasty, even terrifying, and causes us pain. And this relative won't leave. Consolation will never again be possible.

I'm sure Richard had a good idea as to why he was falling, but having refused any further skull or brain scans, he preferred not to know if his cancer had spread and invaded his liver, which led to the wasting, or penetrated his cerebellum, which added to his imbalance. He always preferred to stay ignorant of painful subjects. Imagination supplied the details he wanted me to omit. He trusted his intuition regarding what he shouldn't learn about, and he didn't chide himself for that. He was naturally adept at joining only those conversations that interested him; there were so many topics he wanted to discuss, so why waste time on those he didn't?

He knew it was never going to be safe in his body again. Recovery—being healed, restored, saved—seemed distant. Real helplessness was close by, and the worst part was that he couldn't see anything to do about it. He would be stumbling until he died. Toppled on the floor of his kitchen, he would always be angry—a feeling he'd earned—but also upset and fearful. After he was righted again, he waited, tethered by fear. Fear was a flight from logic and order into illogic and disorder, where nothing made sense and nothing was certain. On the floor, howling, maybe there *were* no words to utter. Not yet sixty years old, he was unlucky. Illness was upon him. Death, which Philip Larkin calls "the solving emptiness that lies just under all we do," was near.

On a Monday, two months after surgery, Leila came to an appointment wearing a black sleeveless T-shirt. Something had shifted. She extended her neck, no longer holding it awkwardly and self-consciously. Her gait had lengthened and returned to the stride I remembered. In the mirror, she pointed out a few landmarks—the contours of her Adam's apple, the boomerang curve of her mandible. Immediately after her operation, a blanket had felt like a steel bar against her throat. Later, the nerves of her neck didn't register—they had gone numb. Now the skin of her neck was sensitive. In bed she could feel the line made by the covers when she pulled them up, but it didn't bother her. Her neck had been the part of her that spoke too loudly, that embarrassed her, but she had forced herself to look away no longer. I wanted to tell her: my grandmother, post-mastectomy, used to stand naked before the mirror, marveling at the sight of her unembellished ribs. A one-legged older man in the locker room of my town's pool would casually, yet boldly, throw a towel over his shoulder or around his waist, trying to agitate and make indignant those witnesses to his exposure. His body was evidence: catastrophe happens. It can happen to you. Here I am. You can go on too.

Leila's scar was now a badge of honor, a sign of battle. It was evocative. She had become interested in her neck, in what she'd been through. She was beginning to learn that in the hierarchy of illness the disfigured are often treated with respect. According to the code of disfigurement, the more gruesome the scar, the higher the rank. If one is young and disfigured, there are bonus points. There are points for bravery. There are points for overcoming danger. Leila had gained respect.

But even such respect can be double-edged. It entails distance across the chasm that opens between the ill and the healthy. What patients trust about me—or any doctor—is the ability to understand that the moment they become ill they are apart, different, separate from the healthy, that relationships have changed and will change again, that life is cruel, but the carnal pleasures of life may still be available.

Leila had regained respect for her body. The smile I knew formed faintly on her face. She told me that she had begun to wonder if she had *ever* had a scarless neck. She almost couldn't remember. Was she born like this? She sometimes held up old photographs to the mirror and tried to compare. "Why don't we keep bad pictures of ourselves?" she asked rhetorically. "They're at least part of the truth." I thought of the servant bathing the old beggar who had returned to Ithaca and suddenly recognizing the scar that Odysseus had gotten during an encounter with a wild boar in his youth. We are recognizable to ourselves and others by scars, old wounds, the damage inflicted by life.

Before surgery and in the early weeks afterward, it was as if Leila had forgotten something: how to be well. Health had slipped her mind for a while. Recovering, the patient has to try to think of health; suddenly it has to be willed, facilitated. In her early postoperative days, this became Leila's job.

"Returning to health is not the same as imagining that nothing's happened," I told her. She could get back to where she was before, but she had lost her implicit sense of security, her sense of trust. How she lived in her body had changed and would continue to change. I appealed to her notions of growing up—she was no longer the way she was at fourteen, was she?

Recovery is a matter of trying to get back, at least partway, to how it used to be. (Tourists call this sense of how it was "home.") "Used to be" is the special past tense of illness. My neck used to be intact—now it looks like half-eaten prey. I used to have no problem with my nudity—now I want no one to look at my ravaged neck. I used to think of my body as a unit, indivisible, the woman who is me—now I don't.

All patients resist the news that it used to be one way but now it's another. The disfigured teach others that their features and form are not inevitable. Five fingers, smooth skin, two breasts, two legs, an unbroken line of nose, all may be temporary. We could lose anything or everything at any time. Pain and terror could come around sooner rather than later. What is good and beautiful doesn't remain unchanged. Disfigurement offers the most literal understanding of loss, of change, of the fragility and vulnerability of the body.

Was it just time that had healed Leila two months after surgery? Had worried self-examination been occluded by other more mundane concerns? "Did she tell you she has a new boyfriend?" her mother asked in my office the last time. *How simple,* I thought. *He doesn't care how she looks.* Leila was wonderful, noble, spectacular, worthy of a young man's love. I was delighted. Too often the disfigured, the ill, feel unloved, and there is never compensation for that. The world of love had invited Leila back in.

Part 4

LONELINESS

Illness involves a specific loneliness, a set of limits and invisible walls surrounding the sudden and incomprehensible crime and betrayal that has occurred to one's body. If illness is a form of travel, it goes, in Oliver Sacks's words, to "a hole in reality itself, a hole in time." Loneliness is a measure of depth, the distance into that hole or chamber where the patient has fallen and is separated from other people. Sacks noticed the hole after he suffered an incapacitating leg injury that tore tendons and nerves, required major surgery, and put him in the hospital for weeks. For him, the hole was soundless, timeless, motionless, and devoid of music, which was what he most yearned for. Patients may turn away from the hole (itself a symbol of death), or they may not notice it for a time because their attentiveness is weakened by pain, or loss, or terror. But when they are in the chamber of loneliness, they get no meaning from intelligence or clear reasoning. Only memory, imagination, and hope provide possible footholds of escape, and even those can be hard to find. With the arrival of loneliness, the patient is *lost*.

Loneliness is intertwined with all illness, from the minor head cold to the terminal disease. Patients experience loneliness whether they are at home or in the hospital, whether their sinusitis came on yesterday or their diarrhea has been relentless for a year, whether they live alone or with a spouse and eleven children. This is not to say that loneliness is one thing to all people. Just as the betrayal of the body takes many shapes—insidious or violent—and the ill body can be experienced as traitorous, disgraceful, struck down, treacherous, cracked open, wrong, broken, or in trouble, there is no single form of loneliness. Indeed, the crime of illness is committed upon those who each have their own natural aptitude and capacity for loneliness. Loneliness has no single meaning because the sick are physiologically unlike each other and have different personal histories and circumstances, as well as different means of coping and supports that can temper suffering. But at some point in any illness, short or lengthy, terminal or transient, the patient who listens closely will hear loneliness echoing as a deep oboe lost somewhere in the vast chamber.

When I began my practice, I cared for Charlie, a tall man, over six-foot-four, with a coarse and richly colored beard that reminded me of parsley. He lost weight at the same rate Richard did until he was 120 pounds, half his former 240. Although I thought of myself as a proficient doctor, persistent and occasionally astute, I had no way to stop Charlie's emaciation and advancing HIV disease. I had diagnosed him, and we both knew the outlook. Charlie came to see me twice a month until he wasn't strong enough to make it in any longer without bringing on more grief than it was worth. At the start of every visit, he would hug me rather than shake my hand. Charlie didn't accept the rigidly conventional handshake. He wanted the tilt and pressure of the whole torso.

I've never liked hugging patients. I believe that medical work demands some sort of distance, or at least the appearance of it. With my patients, I've aimed for friendly restraint, earnestness mixed with readiness. Patients want doctors who are alive and open, transparent and truthful, but they also appreciate that a doctor's distance is a reflection of what is happening within them, the drawing back, the holding off. My guardedness was not ridiculous, as Charlie sometimes suggested, but rather my way of concentrating. Physical remoteness was the price of my attention. I understood that, for some patients, not hugging back could be taken as a sign that I was not devoted to their care. Charlie teased me about my wariness. "When was the last time you cried anyway?" he asked in his soft Southern accent. Then he began to tell me about his symptoms, his grade school friend who had recently come to town, a phone call from a former lover the night before, his favorite 3:00 A.M. television show, the best Bloody Mary in town, his mother in South Carolina. The sick person is allowed to be self-involved, but Charlie had always been accustomed to talking about himself. To him, vulnerability and braggadocio were equally charming.

Charlie had long brown hair, pale green eyes, prominent cheekbones, and a sharp chin. Some days he wore mirrored sunglasses. Whatever assistance he received at home, it was clear that he still picked out his own clothes: green and pink shirts, ties for belts, fur-lined moccasins. His thumbs, double-jointed, dramatically curved toward his wrists when he buttoned his sleeves or gave me a sign that he'd had a good week.

About a year into our time together, he decided he didn't want to be hospitalized ever again. "Why do hospitals have to look like sickrooms?" Charlie asked me rhetorically. Hospitals were dim, underfurnished, sullen, closed-in, and constructed as if no care had been put into their lighting, architecture, or furniture, he said. Off-white and death-colored, there was no coziness in them. To be lodged in a room that gave a sense that someone had died in it, whether a century before or the day before, was not acceptable. He had seen too many of his friends in such rooms, swollen, deformed, devastated by disease. He told me that in the hospital medical care becomes an "ugly pageant" at the expense of sick people.

I also knew that in the hospital Charlie felt cut off, isolated from his friends and family. He told me the doctors were "so damn awkward and so damn formal," and the nurses were justifiably afraid his fluids would infect them. I thought of the bone marrow transplant patients confined to their germ-proofed, claustrophobic rooms on the floor below Charlie's. Their rooms were like tiny boxes, so short on space and air; separated, looking out into the hall through Venetian blinds, patients seemed far away. For Reynolds Price, the hospital was "a marooned island of damaged men and women intent on bringing ourselves to a state of repair that would let us visit the mainland again."

Charlie, at least, could roam the halls. Walking, talking to other patients, he attempted to act as a person and not merely a patient. In his room, getting transfused, he was needy, helpless; outside it, he was again self-reliant, and it was his character, his personality, that determined what came next, how people responded to him, how a room might change when he entered it. As a visitor to other patients, he used his skills as an actor. Depending on whom he was visiting, he was comic or melancholic—an old hand at illness or amused by a new diagnosis, a maniac or a one-man show of slow-motion farce. Charlie played many roles. Back in his room, frustrated and grieving about his condition, Charlie tried to charm the hospital staff and fought the illusion that by sealing off the ill person we can also sequester illness.

Illness had touched every aspect of Charlie's life, so it was not sealed off—it had become him. Although he had been in the hospital plenty of times and he could see many more overnight stays in his future, in one fell swoop he refused them all during this September office visit with me. He didn't want to be detached from everyday living, even as compromised as it was. "Fuck the hospital. No mas," he said. Charlie had a foul mouth; he liked dirty jokes and told them mostly with his hands, delivering the punch lines with great thrusting, obscene movements. I wasn't surprised by his vengeful curse. Each time he was in the hospital, Charlie was assenting to the real possibility that he might not get out again.

Charlie refused to be intimidated by illness. He had no interest in being demoralized. I remembered the last time he'd been in the hospital, the month before, when he'd tried once again to re-create some semblance of his everyday life in this foreign place. Sweating in the August heat, Charlie posted a set of rules over his bed. *Don't lower your voice. Don't feign heartiness. Let me lead. No whining or scared sarcasm. Staring allowed. I get to tell people what to do. All parts of the body are open for discussion. No saying, "Good luck."* Charlie believed that the phrase "Good luck," when used in the hospital, had the same meaning it had when your older brother said it to you before a fight: it was insincere, all theatrics, and inevitably meant that you were going to lose.

Charlie savored the light moments in the hospital, but he didn't do well with the long, futile hours after his visitors left. He didn't like being pensive—"Am I lost in thought, or am I just lost?" he asked me. He grew bitter talking to himself in the gloom, giving the nursing staff nicknames, bad-mouthing the soap operas on TV, smelling the oranges and burnt toast in the hall, being woken by the loud breathing of the patient next door. He didn't want to be just another patient. "As a patient," Reynolds Price wrote during his treatment, "I was in most ways an entirely average floater in the crowded wake of disease and its aftermath. Any special interest I possessed as a patient was the result of the extraordinary size and intractability of my particular cancer." Although Charlie's disease was all too common, I remembered him telling me, early on, "I want my illness report to be on the front page of the paper. I want people hanging on word of my progress, wondering how I am, how long I'll last." Short of this, he didn't want to go to the hospital, and he didn't want me to be his "jailer." He thought of himself as unkillable anyway.

Patients are offered a deal. Doctors will be steady and professional, illness can be managed, and things will go well if patients stay cool and follow advice. That's the agreement. In normal medical exchange, patients are not supposed to refuse care. Patient and doctor are supposed to make a show of solidarity and defiance. Together, they concede little to illness; they loathe illness, and even minor setbacks are outrageous. As a young doctor, I felt jilted by Charlie's decision to stay away from the hospital. I turned away, spurned, deprived of the possibility of purpose. I wanted to be indispensable. Now I never think I'm any more than a steward for the good and bad things that happen to patients. I don't believe that I always know what's best for them.

Some patients have no interest in explaining their decisions, and they offer no openings for discussion or compromise. When Luke first refused surgery, he left most of what he felt unsaid, in large part because he was unaware of it or couldn't name it, but also because terror had blocked access to what he felt. Charlie, refusing future hospitalizations, was more direct: "I'm impossible to please, aren't I?"

After no more than a moment of hesitation, I agreed with Charlie's decision. What happens to people in hospitals, what is done to them, is unimaginable. Sometimes as I drove home in the dark after a day's work I imagined the intercom in all patient rooms repeating: "Your life can end here."

Charlie had a sister who, with the help of her teenage daughter, took good care of him in a three-bedroom cape house on a small lot in a modest suburb. I pictured its ochre walls, the position of the stereo speakers in the living room with the tower of CDs next to it. I could feel the weight of the blanket Charlie wore over his lap when he watched TV in the footed pajamas he adored. That October his sister asked me to visit Charlie, to do a house call; after all, Charlie continued to take his medications and still had nausea and muscle pains, even if he didn't want to be hospitalized and didn't have the energy to make it to my office. She asked me to come by the house more than once. I arranged for a visiting nurse to check on him, to measure his temperature and blood pressure, to draw blood and send me the test results. That was all he needed. "He's had plenty of visitors, but he'd like to see you, if you're not too busy," his sister said on the phone. "Maybe if you drive home this way."

Illness induces, and perhaps demands, loneliness. One is lonely because there is nothing to be done to alter this state. In part, the quality and intensity of the loneliness of illness depend on the patient's attachment to his body, what he needs it for. The loneliness of the ill steelworker differs from the loneliness of the ill writer, but in both cases the body provided a secure base to return to and sickness has caused a violation of that security.

When we are well, our bodies send messages: yearning, hunger, fatigue. We are *sensational*. Our minds interpret these visceral messages—mixing, mingling, complicating, and confusing them—to make us sentient rather than merely sensational. We are used to controlling our bodies: we take them to the gym, where our minds concentrate to lower our heart rates and slow our breathing; we move ourselves along the bowling lane, arms lifting, legs lowering us as the arc of the outstretched ball flows from shoulder to floor, our minds aimed toward the head pin.

The chilling message of illness is that the body has a life of its own. Our minds, we understand most clearly when we are sick, *follow* rather than lead. The body is despotic. During illness—when the body is mutinous, in revolt—it takes us to a country where the children are baldheaded (pediatric oncology) and the adults have tubes to their bladders (urology) or oxygen necklaces (pulmonary). We are led into a new psychic landscape that is barren, prisonlike, a hole. It is a lonely place we don't recognize and have never seen, though somehow we always knew it existed. We can't judge distances, scents, relationships, threats. Illness is a land of tests and trials. Actions are automatic, obligatory, done unquestioningly, yet we are confused. We are ready to give up, to fall apart, but we hang on. With the body's betrayal comes sadness, and sadness is terrifying when it makes perfect sense.

In the chamber of illness, in solitary confinement, the patient relinquishes the powers he normally commands. In good health, he moves in physical and social space. But when illness physiologically limits him in specific ways—he cannot walk, is unable to touch his chin to his chest—he gives up other abilities until his more general powers of doing and going are gone. In my office, a patient sometimes says, "I'm leaving now," but his feet refuse to move. He wants to get away, get out, never come back, but he can't figure out how. Lose even a small amount of bodily function—you can't take a deep breath, or lift an arm—and non-use of the full body may follow. An irrationality, a rival logic to that of healthy thinking, comes into play. Certainty is grounded in the certainty of the body. Health is grounded in the certainty of the body in action. Richard was a man of certainties, but he felt unsteady, he told me, and almost drunk, his head muddled and swimming, as he left his doctor visits. In the midst of a thousand uncertainties, the patient's few convictions are enhanced: *My room is overheated. The food is perilous. No one ever lets me sleep. My doctor is humorless but will do everything to save me. I have every right to be furious. Life is sweet—why can't I have more of it?*

The sick world is limited, contracted. Physical space changes. The sickbed and sickroom become all to the patient, who feels locked inside. He can watch television, but no picture can replace daylight. The sick have no place to go. Right-minded people don't want the sick in their field of vision. Being in bed feels like a continuous state that brings neither rest nor respite. In a windowless cell (and there are hospital rooms with frosted glass or suffocating curtains), the outside world is quickly forgotten. As Brodkey noted, "There is an implacable dissimilarity between the people and events in the active world and the people and events in the grip of medical reality." Space and life are cut off. The sick person's world contracts until he lives in a new half-world.

Charlie's home confinement was self-enforced. His sister's house was safe. There was the familiar certainty of staircase and laundry, of railing and bedroom. He knew the chairs and chores. For Charlie, living with his family was a way of fleeing while staying put. It was a strange indoor island world.

The forced solitude of the quadriplegic, for whom illness never ends, is an extreme example of confinement. Jean-Dominique Bauby, a French journalist, suffered a brain stem stroke and was left bedridden, with only the ability to turn his neck and blink one eye. In a single day, he had lost the many pleasures of the senses—a slice of sausage melting on his tongue, women in flowered sundresses, youths on roller skates, the glass facade of his former workplace, the feel of a shave—but his mind continued to move him around as his body could not. "My old life still burns within me," Bauby wrote, "but more and more of it is reduced to the ashes of memory."

His new life was one of stillness. Bauby watched his visitors enviously as they looked on, not knowing what he thought. He was hungry for human connection despite the toll on his body, the ambush. Devising a means of communication despite these limitations, Bauby had an amanuensis record his "travel notes" in *The Diving Bell and the Butterfly*. "Nothing was missing but me. I was elsewhere." With every memory, he was aware that he was fragile and perishable. How did one ever get used to the idea of illness? Bauby made clear that the imagination had to suffice.

For the patient who is less static, the past has no relevance. He has no goals beyond feeling better. There is dominance of the present. Yet even for those with less severe illness, the feeling of being shut in is extraordinarily intense and, for the most part, unconscious and unrealized. At home in his sickbed, the patient is alone, incommunicado, excommunicated. Often, just when imagination is needed, it recedes, exacerbating the loneliness of illness. It is unnerving; the thought of madness crosses the patient's mind. On the surface, he may seem amiable, even engaged, as others try to reach him, his bereftness as concealed as Bauby's. Loneliness is an inward, spiritual posture. The lonely patient tells himself: *Wait. Be still. Do nothing. Acquiesce. Collect your pride. Endure.*

When the patient relinquishes activity and allows passivity, he is, at first, chagrined. But he quickly comes to think that passivity is the only approach to patienthood. It is as if he is holding still for a grotesque photograph that no one comes to take. Passivity feels like preparation for death, that time when things cease to happen. During illness, something is amiss, deeply the matter, without precedent, without clear meaning, but there is no happening, no clear content; one is simply ill. When any action is useless distraction from the self-absorption of illness, passivity is welcome. Staying still, the patient believes, ensures that nothing worse can happen.

It is immediately clear that the medical world prizes stoicism. Consider the medical trainees who work twenty-four-hour shifts, who come to work sick, sparing colleagues from working extra shifts while jeopardizing their own health and that of patients. Doctors often don't know what to say to patients, and so if patients behave as if nothing is wrong, everyone is thankful (doctor, friends, family), saved from offering faulty and awkward sympathy. The patient who is stoic and silent keeps his doctors from having to breach the submerged sadness of illness. At certain points, solitude seems right to the patient who wants to live in secret. But what are the costs?

There are patients who isolate themselves, who insist on having no visitors, on taking no telephone calls. Who can blame them? But Charlie was not one of them. He was not like Chad, dying of AIDS in Alice Elliot Dark's short story "In the Gloaming." "People who had seen him recently were shocked by his appearance, and rather than having to cheer him up, he felt obliged to comfort them." Yet Chad wanted nothing to do with this. "He had said more than once that he wasn't cut out to be the brave one, the one who would inspire everybody to walk away from a visit with him feeling uplifted, shaking their heads in wonder.... He had liked being handsome and missed it very much." Chad had lost his looks. His flesh was a clock, ticking down. "He went into a self-imposed retreat, complete with a wall of silence and other ascetic practices that kept him busy for several weeks." Solitude may be a symptom of loneliness, but it does not produce the loneliness associated with illness. Loneliness arrives with the body's betrayal and with terror and is augmented by loss. When good health recedes, patients immediately become psychically unavailable.

Only rarely is a sick person physically isolated, as with the self-directed exile of a patient like Chad or the quarantine of the saddest and most infectious cases. Paradoxically, the opposite is more often true: the sick see more people than they ever did when they were well. It is common for patients to be lonely, however, even when they are surrounded by people. Think of an intensive care unit: what other environment could be so lonely and so incredibly crowded at the same time?

As the ill person becomes intimate with his new, unfamiliar body—his new accompanist—he doesn't like what he hears. Harold Brodkey, who discovered he had AIDS at the age of sixty when he developed a sudden and severe pneumonia, wrote, "I felt myself to be thoroughly repellent. I had disowned my body and was mostly pain and odors, halting speech and a sick man's glances." Brodkey could no longer remember the body he used to feel, "that odd, flexible, long-limbed extent of reliability" when "all the tubes of sensation flashed a little in a silent fusillade, and in private, one stretched in a courtship display." When patients are sick, they pay attention to the body in despair, its new rhythms, noises, quirks, weakness, off-colors, soreness, signals they don't know how to interpret as they do in their usual state. Disgust and revulsion deepen a patient's loneliness. Brodkey, in his inimitable way, called himself "death meat."

Within the deep chamber of biological identity, the patient becomes a new "self." His old identity has been spoiled. Don't most patients say, when they get better, "I feel like I'm myself again"? This unexpected intimacy with a sick self demands a treacherous new relationship with the body and affects all relationships. Sick people pay less attention to others. The body's betrayal dispossesses them of intimacy with loved ones so that room can be made for greater self-involvement. In this way, patients distance themselves from family and friends—those who are well. The sick maintain some loyalty to the outside world, but the real news, as Brodkey wrote, comes from inside.

Illness has certain properties that feed and amplify loneliness. It is juvenilizing—illness makes a thirty-four-year-old man like Charlie wear PJs with scuffy, rubberized feet. It is understood that illness returns one to the early helplessness of childhood. In illness and childhood, the individual feels vulnerable to attack; indeed, illness is a sign that he has already been attacked. A patient is *unprotected*: he is subject to the appetite and whims of the illness that inhabits him; he feels persecuted. The ill person is lonely because there is *no one to blame* but himself; he is failing himself. Like the child who will do anything to resist self-blame, the patient can sometimes fight back, at least against minor illness, and in the most childish form of counterattack he blames others: "*You* gave me this cold."

Protection is needed at these moments, and sometimes illness is accompanied by its own self-protection. During illness, a patient often doesn't feel—or at least doesn't admit that he feels—the full extent of his problems. He ignores his problems out of fear or hopelessness. His mind tries to trick him into believing that if he is cut off from what he feels, it doesn't exist. In this way, he protects himself from the next betrayal. For a while, the patient fights back, but at the same time he already knows he has lost and surrendered, and it is in these moments of giving up that he experiences loneliness most acutely.

"I don't remember whether I was afraid of this test I'd never heard of," Lucy Grealy wrote, "but when my parents said, 'Well then, we'll be off,' I looked at them panic-stricken and asked, 'Aren't you going to stay with me?'" Left in the hospital to cope alone, "I felt my face flush. Things seemed to rush at me as if I were the focal point of some unseeable camera's close-up. Immediately I regretted all my assumptions.... It was the moment when I understood unequivocally: I was in this alone." In later years, Grealy could summon this loneliness; her earliest treatments made a lasting impression. She grew weak from certain memories and smells. Memories of other illnesses return when patients are sick. Sometimes ordinary objects in the world assume sickening old meanings. When Grealy saw a napkin, it reminded her of surgery. Another person's glimpse of a metal bowl recalls whole days of vomiting. Such memories, in combination with one's current illness, make a continuum of illness, of wet, foul, sticky, shapeless, anguished days.

When we're young, we don't think of minutes passing, but to sick people time is critical, and if illness is not life-threatening, time is what they are always aware of. Fluid and unpredictable, time is also very confusing. And it's not always passing in slow motion. Sometimes illness is like a sled, as Brodkey described it, and there's a downhill excitement to the run. For the postoperative patient eager to get home, every second spent waiting for the final intravenous line to be withdrawn feels like an indulgence. But more often the hours seem endless. The patient in traction with a broken hip finds himself parked, stashed, left behind, and ignored. Time can be tectonically slow, dense, broken into sections, gashed and fissured. But at the bath hour, there is a gentleness and spareness to the moments. Television, that time neutralizer, plays incessantly. The mind is present, then wanders. Hospital nights are not nights filled with sleep, making the day go on for a full twenty-four hours. Clocks are everywhere. As the memoirist and bone marrow transplant recipient David Biro wrote, "The clock on the wall is a stagnant pond; algae is accumulating on the top curve. I follow the second hand. It is the only thing moving. I want to push it along."

Illness feels irrevocable, so the patient imagines it will last forever, and in this way sickness changes time. When a patient passes through illness, time no longer feels abundant, and he is no longer careless with it. Time also offers patients something on which to pin hopes (*I'll be better tomorrow*). Still, how can the sick escape the feeling of hours dragging by?

They receive company. While the arrival of hospital visitors can be joyous, visiting hours also generate a certain tension. Patients know that these hours are limited and that they will be left again with their failing bodies. Visits remind patients that sickness cannot be shared, only briefly observed. It is an individual sport. Loneliness, like pain, cannot even be really explained. The process of illness is essentially private; it takes place day and night in a solitary chamber that offers no span or reach, no range of vision. Illness, by its very nature, excludes others.

So why are visiting hours on a regular hospital ward unsatisfying? In part, it's the forced politeness of the visitors. Conversation is awkward and clumsy, owing to its unnatural delicacy. Politeness is a grim signal to a patient, as Charlie made clear with the sign above his bed. He read worry and fatigue on the faces of friends and family members and realized that his pre-illness relationships with these visitors had shifted. In their attempts to care by being overly sweet, the well made him feel separate and different, exacerbating the feeling he'd had when his body deserted him. He wanted to be the same; they treated him differently. Visitors bring flowers and look around for a place to hide their concern. Charlie had suddenly understood the term "visiting": living apart, he was in one place, visitors arrived from another, and in the brief interlude of "visiting hours," they could not know where he had come from.

On the other hand, conversations with a roommate in a two-person hospital room are under the patient's control. This is why the bond between sick roommates is often immediate. Part of a patient's loneliness is that he loses a sense of people except those who are like him—sick. A hospital roommate can share in the conspiracy of illness, the overthrow and revolution that is the body's betrayal. A doctor stepping into a two-person room to see his patient sometimes has to admit he is treating a pair. As the medical interview proceeds, the doctor hears from beyond the curtain: "You forgot to ask him... Tell him about the... That's not what you said last night." Hospital roommates who hit it off consult on everything. This is what happens naturally when any two people share a bathroom. They talk about soap and cigars and how much they had riding on last Sunday's Patriots game and anniversaries and high heels and ice-fishing and recipes and divorce. Roommates don't mind crying in front of each other because they silently agree on the facts of hospital life: pleasure leaves too early, and trouble returns too quickly. In parallel beds, patients rattle like two tugboats beside each other, pulling the tonnage of disease, embarked on similar routes. A hospital roommate is someone whose life is as difficult as the patient's own. Of course, sometimes this similarity is frightening and leads to silence. But more often patients trade secrets—the alcohol under the sink, the son's bedwetting, the years of couples counseling, the handgun in the bedroom drawer—to find equilibrium and admit to the distance from life before illness. Trading secrets is the surest sign of a defense against loneliness. A secret is a promised connection into the future (one that seems lost during illness); implied in any secret is the promise that you will not be abandoned by the holder of your private knowledge, and this promise offers the patient another means of protection: the vitality of the information he has entrusted to someone else. Secrets are alive.

There is nothing lonelier than a single room. Things drift into the illness chamber from the outer world. According to Brodkey, "The television is a window, the telephone a murmurous keyhole." It's bad enough to eat dinner alone at a restaurant when you're healthy. Try three meals a day served on a plastic tray that's been reheated in the hall and left on a lever-elevated, rolling table that won't quite rise over the tops of your knees.

Despite his sister's calls, I didn't visit Charlie. His house wasn't on my way home, and I didn't really believe he wanted a visit from me unless he required a specific treatment. His sister was just being polite. Charlie was rarely alone in any physical sense. He had lots of friends over at the house whenever he called to ask me for a refill or to notify me of a new symptom. Unlike Brodkey, who wrote, "I have lost my sense of people ... I like to be alone, me and the walls. Do what I do, think what I think and to hell with the rest of it," Charlie was popular and gregarious. Members of his extended family *extended* themselves and visited from far away. He had loved ones who helped him clean himself, who rubbed his aches, who trimmed his earthy beard, and who stroked his beautiful shoulder-length brown hair. There was never a time when his sister called to check on something—a prescription refill, a side effect—that he was alone. I was able to assure myself that he was not in pain, that he was taking the proper medications for the chronic condition I was treating, that the visiting nurse was doing her job. If I visited him, I told myself, I would not really be coming as a doctor, even if I did carry a black bag. I didn't need to check his wounds, or give an injection, or even examine him. I would arrive differently from any other visitor: I would be awkward, halting, and inadequate. His disease was advancing, and there was nothing more I could do to slow it. Maybe that's why I didn't go: I was a reminder of failure, his and my own.

Loneliness is never the chief complaint of patients. No patient says, "My loneliness began three days ago," or, "I feel lonely *here*." This is partly because loneliness is often not a strong or clear state of feeling, like disgust or panic or rage. It is harder to identify and name. In addition, loneliness is often not apparent to a patient or caregiver until after the more pointed experiences of terror and pain are dealt with. Yet loneliness, like pain, is unshareable and invisible, and when sensed by the patient, it seems like a private fiction (*Is this what I'm really feeling*?). Loneliness is hidden. More public are the shame and apology, the fear, the helplessness and dishonor of dependence that come with the diminishment of health. Arriving soon after the chief complaint—the shortness of breath for four days, the vomiting that began last weekend—loneliness is the universal comorbidity of illness. But the word "lonely" isn't heard often in hospitals; more commonly, one hears doctors talking about patients being withdrawn, agitated, or depressed.

The *Oxford English Dictionary* definition of loneliness is "dejection because of want of company, sadness at the thought of it, a feeling of solitariness." Such aloneness is part of the loneliness of illness. But the loneliness of illness is not a sensation like the heartbreak of lost love, which is associated with an absence; nor is it exactly the I-have-no-place-in-the-world feeling of an orphan. There is apartness to illness, and the image of illness as a chamber implies a spatial element, a sense of travel.

Distance can obviously be not merely physical but psychological as well. When a person is sick, it is as if he's gone away. As friends and families and lovers retreat (sex, which is an attempt to escape loneliness, disappears during illness), the patient withdraws from them too. The sick person lives at the vanishing point, that great invention of the Renaissance by which artists measure distance and by which doctors measure the approach of death.

Death. Isn't it a tiny, passing worry even during the first day of that cough and runny nose? Is this cold the beginning of pneumonia, or avian flu? When life's stability is threatened, we think automatically of death. Our bodies, the great comfort and pleasure machines we don't think about until something goes wrong, have deserted us. And instinctively, sometimes immediately, we sense, no matter how serious or minor the illness, that we have reached the start of the vanishing point where life disappears into the distance, where life separates from death. Death is the ultimate and eternal loneliness; illness always brings with it this shadowy fear, even if only fleetingly. Rationality has little to do with it. When you are in bed at home, immobilized by pillows and blankets and a thermometer in your mouth, and you hear your children downstairs getting breakfast for themselves and your wife on the phone previewing her day with her friend, you think: *So this is what life will be like without me. This is what life will look like in this house when I'm dead. Things can and will go on.*

There is no body-memory of health. Although we can remember what we *did* when we were well—raked, shoveled, blew on a French horn, jumped the KEEP OFF THE GRASS chain—the memory of how we *felt* doing it either no longer exists or is infinitely far away. Some part of illness is doubting that the sensation of being well ever really existed and doubting that it will ever return. The sick person is thereby cut off from himself or his remembered self, and the gap seems absolute.

"I had felt my aloneness on the mountain as being, in a sense, almost sadder than death," wrote Oliver Sacks about finding himself stranded and immobile after a muscle-ripping fall on a Norwegian alpine trail. Desolate and deserted, loneliness was like death in its stripping away of communication and reassurance and hope.

I've been assured by many psychiatrists that not all people with terminal illness are depressed. Early in my career, I didn't believe them, even though in my limited experience not everyone with end-stage cancer was miserable, apathetic, or detached. I understood that at the end of life any philosopher could see that the mind and the body are separate, that humans are everlastingly resilient, and that the loss of the body can be compensated for in some ways. But still, I argued, those approaching death had to be depressed—how could they not be? Now I understand that it's not depression that is omnipresent in the terminally ill; it's an unfathomable loneliness that may or may not overlap with depression and that I certainly mistook for depression when I was younger. What is this end-of-life loneliness like? I can't really grasp it, not the doctor in me, not the writer in me, not the middle-aged man either. The only way I can approach this loneliness is to think of it in fairy-tale terms: the loneliness of being in a forest at night, of being watched and perhaps soon devoured.

I told myself that when I had a few open hours, I would drive out to Charlie's place and back for a visit. I first considered going out to see him when he said he couldn't make it into the office, when he couldn't walk far anymore and his sister didn't want to import strong men to put him into a car or a wheelchair. But I didn't. I sent the visiting nurse. I never visited him, but then again, he never asked. Only his sister—speaking for Charlie, she said—asked me to come over.

I didn't think Charlie was about to die. If he'd asked me, I would have predicted he had at least another ten months in him. I was planning to visit when my schedule cleared out a bit. This is what I told myself when I heard that he'd died on Christmas at eighty-four pounds, "after a long illness," as the obituary put it, at home, at the age of thirty-four.

Charlie once suggested that being sick is a kind of preparation for being old. There are bodily rearrangements. There are lousy smells, foul fluids, and decay. There's fatigue around the eyes, empty-headedness, resignation. His sister told me that in the last months of his life Charlie's friends believed he was slightly out of it, as if he had already renounced the world and therefore had stopped bothering to make sense of what he said; worse yet, Charlie felt that he had little right to attention, a real change for him. He even tried to wave his sister away, faintly embarrassed that he needed someone to wash his face, to shut off his light. He felt left out, a straggler; he crept along, and no one wanted to wait. The fear of sudden disappearance, death, lurked. There was the lingering question of what his life was worth anymore.

Like Luke, Charlie was bothered by these thoughts when he was awake at night while others slept. When I spoke with him on the phone, he mentioned that he was often up the whole night wishing he weren't. He spoke with a kind of detached wonderment. Sometimes he read, but he was fascinated by infomercials and could watch the selling of a food processor or a clothes-cleaning vacuum or a skin-care cream and wait for the light to come. He had lost the rhythm of rest and activity, of plans and expectations. The sociologist Arthur Frank could have been describing Charlie's condition when he wrote about his own cancer, "I was neither daily nor nocturnal, but suspended outside the limits of either existence. I was neither functionally present nor accountably absent. I lived my life out of place."

How does a doctor *know* that an ill patient is experiencing loneliness? Is loneliness susceptible to scientific scrutiny? Is this deep experience, transient and subjective, meaningfully measurable? The scientific questions that follow might be: Does loneliness have attendant morbidity, that is, does it exacerbate the underlying illness? If so, is there anything doctors can do to ameliorate the condition, lessening its ill effects?

I believe that nothing other than getting better can completely undo such loneliness. If doctors could come up with an intervention prior to the complete clinical resolution of an illness, it would still not cure a patient's loneliness because a patient can never be wholly convinced he is safe and secure until he is well again, and if he knows he will never be well again, he must be unbearably lonely.

If a patient cannot even say to his body, "Get better," what can he ask of it? He wishes he could have an existence that has nothing to do with his contingent, mortal body. This is dissociative; he feels keenly the great expanse, the chasmal space not only between himself and visitors but between his private, inner travels and the combination of unpleasant sensations that is sickness. Leila knew better than most her body's natural shortcomings. But when the surgeons came to work on her neck—one small piece of her—she tried to dissociate her body from the rest of her life. This doubleness was expressed in her belief that her neck was part of her, but not *her*.

When a person with chronic illness is out and about, no longer in the hospital or in bed where relatives can gather and offer sympathy, no longer wearing a johnnie or pajamas but shopping for groceries, passing papers on a mortgage, or drinking coffee at the office, she is carrying a secret identity. Who around her knows that only last week she completed radiation treatments or began chemotherapy? Who knows about the ceaseless foot pain or the terror of upcoming surgery, about the itchy scar under the blouse? Who at work knows the inconveniences of her illness, the series of discomforts, each tolerable, but one after another allowing no rest? She doesn't "look" sick, so she must not be. The fact that she can pass among strangers while holding this news close, revealing nothing to them, exacerbates the loneliness of her illness. Somehow she manages the grind of daily life while illness looms over everything, shadowing, trivializing, obscuring what must be attended to. Patients become so used to going it alone that they develop defenses, evasions, smart-aleck remarks to release them from the obligation to engage in the subject of illness when it arises.

Not all loneliness is alike. It can't be, because it includes lost dreams, an imagined future, and what still might be achieved. This makes it unlikely that a single remedy will be found.

So what might help? What does a sick person want from his caregivers and his doctors?

Patients want their doctors' knowledge of illness. Patients are constantly reminded of how little they know about how their body works. Lucy Grealy described illness in her memoir as a movie in which "something important had been revealed in the opening sequence, some essential knowledge everyone else was privy to that was being kept from me." Patients want the opening sequence. Patients want doctors to say: "I found nothing that surprised me, nothing that told me anything I didn't already know. I understand perfectly the chemistry set of your body." Most patients quickly pass over the first question, the one with no satisfying existential answer—Why? Why me?—and instead ask doctors for details, the trajectory, the plan, the system, the mode of counterattack; they want to know whether or not they will die.

Patients depend on doctors for approval and disapproval, at least when calculating personal blame for illness. Doctors can say: It wasn't your fault. You're a good man. You matter. Yet I can tell that when I say these things patients don't quite believe me. I need to repeat myself two or three times. I always underestimate the effect of this repetition on a patient distracted by fear. Doctors can't always promise to make things all right, or to show patients it isn't so bad, or to cheer them up. That approach is too loaded with lies and misstatements. But it's important for us to say, "You're not quite dead," even if the response is Brodkey's, who told his doctor upon leaving the hospital, "You saved my life. Of course that's not saying much—the whole thing is a mess."

A patient wants his doctor's strength. He is too weak to garner his own. His doctor is his means of fighting back. He thinks of doctors as strong, wise, and compassionate, as the ones in charge of doling out surprises. Doctors bring jolts, blasts of energy, animation, information. Medical training is an exercise in gathering strength. For medical students, part of this strength-work is purely physical—long hours laboring through fatigue. But another part of being a student for four years, and then an intern and resident for years more, is trying to deal with the sheer tension and density of emotion. If, every day, these trainees meet each patient with undefended sincerity, soon enough they will be young doctors who are seriously conscientious, wise, and strong.

A doctor's strength determines a patient's protection. Doctors need to understand that the loneliness of illness is a state of loss and surrender, and they must learn how to protect the sick and fight for them, even if they can't rescue them. Doctors must lend strength in the form of optimism. Optimism can mock death and weakness.

Patients *need* their doctors, and if their doctors are inaccessible, if they don't permit the sick to talk about what they're thinking about, the loneliness of the sick is only exacerbated. To be strong is to inspire patients to hope, even if the hope does not reasonably involve a cure. Here's a secret about doctors: they're all optimists, and as optimists they must be in denial about lots of things. Hope comes so easily to doctors that sometimes they offer too much. Patients, even the loneliest, are mad with hope, and doctors easily match their patients' hopes because their eagerness for improvement verges on the idiotic.

The memoirist Anatole Broyard, weakened from prostate cancer that had spread to his bones, wanted his doctor to stir his flickering ego, to remind him of life beyond illness. He wrote that he wanted a doctor who was "intense and willful enough to prevail over something powerful and demonic like illness." He wanted a doctor whose charm consisted of more than accuracy. Broyard wanted inspiration. He wanted an emotional leader who radiated calm, did not overreact, refused doom, and offered an upbeat judgment.

Small gestures are enormously comforting. The doctor who sits before he speaks. The doctor who offers a cup of coffee, an indication that the world continues. But what patients are hoping for most is the *talk*, because the attachment it creates is a way of fighting against loneliness. They want to feel connected and cared for. What they want is fragile and rather private. It is an intimacy they want, like the intimacy of light as it slips down over the edge of the bay, that closing distance.

Turning to religion not only provides consolation for some patients but is also a way of disclaiming the world's danger. Too often, conversations with doctors are difficult and unfulfilling and weakening, so a relationship with God is more predictable and stable. Alone in bed, in the dim, credible light, prayer provides the mind with a reliable presence. The patient prays when he needs a listener, Reynolds Price noted. He doesn't know who is listening, but he presents his needs and hopes, and whenever consolation arrives, it comes not from an answer but from being heard. Religion, newly found or rediscovered, reinforces the patient's new awareness that he has moved from a trivial plane to one that is tragic or absolute, and it confirms that he is valuable enough to protect.

Charlie used to remind me that he was worth protecting, especially when he felt that he was unprotected and running out of time. "Don't just stand there," he would say when I entered his hospital room. "Come in, it's okay. I'm alone, except for my AIDS. It's always here. Now give me a hug. I'm just teasing you. Or at least sit down, I don't look that bad yet." Every patient wants his doctor's time, which is another way of saying he wants to share some of his time. Sometimes my patients want to tell me everything. Charlie wanted me to know how it hurt when I took his blood on fifteen consecutive days. He wanted me to know how it was to be naked under a thin, backless hospital gown all day. He wanted me to know how claustrophobic the MRI tube was. He wanted me to appreciate that the tests I sent him for were a form of calculated torture.

He wanted me to know how much a catheter hurt. He wanted me to know how he'd have given anything to feel fine again but sometimes he just wanted to be dead.

Of course, I know all of this, but that's not the point; the point is that the patient gets to explain it to me, and that takes time. The patient embellishes and brags because he loves a good tale. Charlie's favorite was about the young woman pushing his stretcher to the MRI area who got lost in the tunnels under our hospital for an hour and ended up rolling him into the morgue—"I think she was dropping me a hint." But the patient also needs me to know that I'll never really understand, and he wants me to know that *he* will never really understand what illness is like either, even as he lives it and tries to speak of it. Misery loves company, Charlie told me, but only good company.

Why do patients like medical students better than residents and better than even senior doctors? Because students seem to have time. Or they take the time. When a patient tells a doctor about himself, it is the beginning of escaping loneliness and shortening the distance he's moved into the chamber of illness. But sometimes it is a widening effort.

Patients are generally forgiving of doctors, even those who seem uncomfortable and are hard to make contact with, and even those who have no idea of the importance of the moments they're passing through. But patients are also on guard because they are vulnerable. When humans need other humans, they watch closely, judging who will come through. A disturbing glance, a sabotaging word, the way a pill is presented, gently or brusquely, can significantly disrupt a world that already feels dangerous.

Sometimes patients are looking for simple advice. Other times they just want to speak with a doctor to no specific end, which is not so simple. When I ask, "How are you?" patients take me literally. They want to give me a true report of their inner lives. To patients, the question "How are you?" has an existential cast. They want to talk, which means they want someone to know that they are alive, that they are still here.

When everything has gone wrong, my patients need me to remind them of what is right; the right things are what they associate with me. Often patients go blank when I arrive. They are unable to remember what they want to ask, they respond incoherently, or their thoughts arrive out of sequence. I am the source of news, of action. I am the embodiment of the chance of winning. I represent respect in the face of humiliation and indignity. When I leave, they may not be strong enough even to show gratitude.

Sometimes the sick are loud in their loneliness: they question everything, they spit and fight, they laugh outrageously, they want a different kind of interaction, and they come to me. Sometimes it's not talk that matters but touch. The loneliness of illness, after all, often involves a loss of physical force and connection. An embrace is worth a great deal and is a kind of language. But it's not everyone's style, as Charlie teased me when I went stiff in his embrace. Sometimes lonely patients are quiet. They are quiet in order to maintain their calm. They believe in this way they can repair themselves. They want others to come to them.

When Charlie was alive, Richard was too. I cared for Charlie in the slick, cool, professional way I knew then, giving him referrals early on, calling in medications, sitting back, helping. But I didn't drive the ten miles to his house. His sister had always said on the phone, "Charlie was asking for you," but he never got on the line and asked me to visit. In retrospect, maybe I was simply, vainly, waiting for *his* invitation rather than his sister's. I could read my inaction as indifference or laziness. Or maybe I didn't understand that Charlie was lonely—I was unwittingly missing this diagnosis. Maybe I had been unable to diagnose his loneliness because during his final office visits I had been in the presence of loneliness without being able to name it, or look at it, or accept it.

But even this seems a simplification and not quite right. I *had* some sense he was lonely. I actually held myself away from Charlie. Being in the presence of his loneliness was like my first look at a cadaver. I looked, and then I looked away. This looking away was instinctual. In part it was cowardice, but also we are instinctually opposed to abandonment, and this is what we sense when we first look at the dead, and at the lonely. Like the dead, the lonely person has been defeated. We sense that he has had everything taken away.

Some piece of me recognized this, and it bothered me. I used it as a silent excuse to stay away, and in doing so, I failed

Charlie. I never responded to his loneliness, and nothing deepens loneliness more than needing someone who doesn't respond.

It was probably the experience of writing novels that finally allowed me to recognize my failure. Every character in a novel is an extension of the author, and trying to understand the range of my characters' emotions allowed me to internalize some of them. The clearest way to make sense of others is through a sense of self; this is at the root of novel-writing, but it should also be at the core of medicine. How can a doctor understand the terror, losses, or loneliness of a patient if he can't imagine it for himself?

I like to imagine that there was a period of time near the end when, every morning, Charlie's eyes still slid open optimistically. When he didn't know how many days he had been in bed, in that utterly remote region of illness, or how many more it would take to get back home. When, if but for a moment, he would forgive himself for getting into this mess. Morning would have induced the illusion of hope and renewal, but during the day Charlie must have thought of the essential foreignness of family and friends. It pained me to think he would experience the dire sensation of being too far from home, at the very edge of existence. Night would bring anxiety and rebuke, a reconsideration of his position, misgiving and doubt as to where he'd gotten himself, as if he'd had a choice. He would call up every bad scenario, every person he'd met who was now dead, people he would have liked to have known better—the landlady at the first apartment he lived in alone, the guy with cowboy boots at the Laundromat whom he talked to every Wednesday night when he was new to town. Charlie probably thought of every other place he had ever visited that would be infinitely more desirable than where he was just then. Even as he closed his eyes, he probably thought that if he willed it or wished hard enough, he might be able to escape.

I imagined all of this, but during my final days with Richard, I remembered Charlie saying, "It's so much better to know that something will end. Tonight I'll accept the help of a sleeping pill, and tomorrow morning will be different. Not better, but it won't be today."

I didn't recognize then that arriving at Charlie's house as his doctor would have meant I was *not* just another visitor. If I had known then what I know now—what I learned from Richard's death, from the writing of this book—I could have leaned in close, crouched to listen to his chest, looked him straight in the face. Even if I'd had no medication to give him, my conversations with Charlie would have been different from those he had with anyone else. I had looked into the chamber of illness many times before, and I could have told him what I'd seen. I could have been the visitor who helped him figure out his position in the chamber, who gave him a different view of himself. I was the visitor who did not have to participate in the conspiracy of reassurance. I could have hugged him and, with my body, dispelled some of his isolation, allowing him to transfer some of his misery to me.

Patients, by virtue of being ill, have the right to ask too much of doctors. Doctors are used to it. I am used to putting my moods aside. Sometimes, if I'm lucky, in return I get to do something meaningful: I get to watch hope and pleasure start up again. I get to see lipstick reapplied for the first time in weeks, wallets stuck back into pockets; the beads massaged on an old bracelet that no longer signifies misfortune are once again charms. The body is redeemed. The chamber opens.

Epilogue

"I have never been anywhere but sick," wrote Flannery O'Connor, who developed lupus, a chronic illness, as a young woman. "In a sense, sickness is a place, more instructive than a long trip to Europe, and it's always a place where there is no company, where nobody can follow."

The living room where Richard lay emaciated in his last months was like a deserted museum, and he was the remaining exhibit. White sheets, metal bars, science and industry, anonymous. The atmosphere was that of a museum's abeyance, the limbo of ritual observation. Sunlight and memory. Everything in the world that mattered and was meaningful was before me. Seeing Richard bed-bound took me to my professional, mental, and emotional limits.

Illness provided its own atmosphere. It was not a matter of place but of a world of diminishing returns. Richard's eyelids fluttered and closed. The quick strength of his body was gone. His head was a weight in my hands. I inhaled the odors of latex and urine and Desitin. I went around his hospital bed adjusting the blankets, and I seemed to drift, waiting for something restorative.

Each time I visited I was taken aback. For the longest time, I couldn't stop looking at him, as though he were a sleeping baby. I wanted to tell him all my secrets; I wanted him to tell me his. I tried hard to retain the sound of his voice. If he'd been awake, he would have shouted, "Stop looking at me like that. What do you want from me?" But he was in repose, wholly unaware, dying. Early after diagnosis, some patients say the word "cancer" as if it is all you need to know about them, as if it is all they *did*. Now cancer was indeed all that Richard did.

I remembered how Richard had been hospitalized two days after that first fall. He had blood drawn in the emergency room, where his surgeon had directed us first, and a few hours after he got up to his room, a nurse came in ready to draw his blood again. Richard refused. "I gave some downstairs, and they said they wouldn't be taking any more until tomorrow." "Your doctor wanted to add a few more studies he thought were important," she said. "He can add them tomorrow, or you can figure out how to use the blood you already have." Richard had always resisted discomfort, but a needle in his arm was no big deal. This time he wanted to hold firm because not doing so was a form of giving up. He did not want his body to be taken over.

Lying on the hospital bed my sister had installed in the living room, Richard had a sleep expression that was poised somewhere between resistance and powerlessness. His lips curled back to reveal root-exposed, yellowing teeth. His face was merely a ruined likeness of how I thought of him. I excused all the times I'd seen disappointment on that face, all the times I'd indulged his foolishness. I thought of all the kind things he'd said about me over the years, mostly to other people, to my niece, my sister, and her friends. Whatever good and important traits he thought I had, I'd tried to maintain. I'd learned a great deal from him about what was beautiful, what was acceptable, what was unacceptable—his personal aesthetic.

My sister was his minute-to-minute caregiver, with intervals of relief from her daughter, who came into town most weeks. Closing down, Richard seemed thankless. Over the years, he had never shown gratitude, or if he did, it was in indirect ways: interrupting his own work when you arrived, offering you praise on your work, unexpectedly inviting you to join him on a trip to a gallery. The sick, by allowing others to help, make their helpers feel good. But now, when he needed it most, Richard found help irritating or unsatisfactory. Maybe this was a sign that death was approaching, that despair was overwhelming, that he was stunned. The cats came out of hiding and lay beside him, untended. My sister, with the exaggerated sense of necessary confidence, ignored his refusals and engineered his pain medications. She didn't ask his permission, and she overrode his prickliness.

Whatever lay ahead was inescapable. Medical care involves practical questions that need resolution. Illness is realism, a kick, a blow. But my medical mind was gone; my higher level of reason had withdrawn. I sat at the end of his bed and simply repeated: "Don't die. Don't die." Richard was traveling, far away. Only years later did I read Arthur Frank's dictum that what the patient needs most is "a sense that many others, more than you can think of, care deeply that you live." I hoped this need had been met for Richard.

I remembered how, in the beginning, years before, Richard had accepted his diagnosis uneasily. Details from that time bled through my shut-down capacities. He understood immediately and naturally that there were things out of his control that could and would hurt him. He must have felt connected to his cancer in some aesthetic way. Cancer was wild, disordered. He collected Chinese rocks and treasured their sinuosity, the way they were gnarled and wrinkled and organic. A worm-eaten wooden beam was evocative to Richard. He appreciated randomness and chaos in art. Associations between forms were always clear to him. His fame as an artist derived from his ability to transform a natural object into art and back again. Richard had as active a relationship with his rocks as he did with his cancer. He was an expert in the body; as a young artist, he had worked in anatomy labs on cadavers to learn the shapes of muscles and bones. How his cancer fit into his skull (CT scans fascinated him) was a spatial problem, and manipulations of scale and tiny openings and magical space, the essential problems of sculpture, had preoccupied him since boyhood.

His insight into his condition was grim and luminous and freeing. In the end, he refused the ritual of chemotherapy, which conceivably could have offered some late control of his cancer. Why? He correctly heard the low odds of success. He heard the evidence presented and never fantasized a cure. He understood the terrible momentum of his disease. Richard was emotionally unafraid. Perhaps that is partly why his illness hit me so hard. He raged and confronted, he was cruel and unanswerable, and so, triumphant. He understood the physical threat of treatment and had a distaste for the inevitable side effects. He enjoyed rebuking the doctors and their understatements of misery. He didn't want to be exhausted near the end, didn't want his interest and energy taken up by near-worthless medications. He never lost interest in others, despite wearying days, bodily attrition, and the natural sorrow of separation. There were a thousand awful ways to die, and he didn't want to think of any of them. He wanted to keep busy, because if he had things happening, he could hold off death forever. He'd always been rebellious and didn't care to do what others readily accepted. He'd led a good life.

When he first climbed into the hospital bed in his home, he cared and worried and laughed over the ridiculous state of his body. He was not timid to ask for what he wanted—morphine, sips of water, the television turned louder, an extra blanket. He invited his doctors to the house, but they kept their professional distance, and he didn't blame them. Richard had no stories to tell of doctors who didn't show concern or listen; now he was happy to have the hospice nurses, who remained calm whatever the anguish. He was not afraid to face his brother, who came to town to reconcile after years of bad feelings. They spent the afternoon sharing tales of good and bad luck, trying not to inflict further harm on each other. After his brother left, Richard decided he would accept visits only from those who, as Reynolds Price wrote, "ask few questions, give no predictions, and make no demands."

He was willing to accept dying, but he didn't want it to be bad. Alive in his mind, he hung on, and it got bad. He was catheterized, infected from his pressure ulcers, intermittently delirious, and alone in his confusions on a single bed.

He never slept deeply. His feet never settled. He was unreposeful, as if he had a grudge against rest. I remembered from the nights I spent as a medical trainee on the hospital wards that when a patient sleeps peacefully, he contributes to a doctor's peacefulness. Richard had never had an interest in other people's peacefulness.

The pressure to say the right thing was enormous, and I was consumed in thought as to what it might be if he asked me a question. I felt the same as I had in the car during that first drive home from Johnnie's Luncheonette. What did he expect from me? Knowledge, understanding, admission of nonunderstanding, affection, truth, a miracle? As his cancer progressed I was alarmed, and it was tricky not to alarm him too. He had a fatal illness, and I was dizzy with this fact closing over me. I had no rigorous distance. But my feelings were intense, and I was troubled, unable to dissemble.

Since he had never believed that I really *was* a doctor, or that I wanted to be a doctor, or that I enjoyed my work (he offered all three of these descriptions at times when he introduced me to his friends), when he opened his tired eyes and looked at me, Richard just saw me as his little brother-in-law. As the doctor of a patient in his condition, I might have retreated—to other duties, to other patients, to the possibility of success elsewhere. My acceptance of the way he had always viewed me relieved me of my worries because I could respond to him as his lifelong fan and relative and not as his doctor. As his doctor, I would have tried to stay self-possessed in order to maintain my own perspective. At his bedside, I could surrender realistic caution and still think of him as everything he no longer was.

Thinking of all I might say to him if he were alive today, I understand that difficult feelings—the four at the heart of this book: betrayal, terror, loss, and loneliness—are the basic, invisible facts of illness. What doctors can do is provide a reservoir of shareable examples of these feelings that they have absorbed from immersion in the lives of other patients. We have learned not to worry inordinately about patients; if we did, we couldn't manage. Swamped by worry, we would be useless. Instead of worrying, we try to help, encouraging the patient to call forth his mysterious, inward responses to his difficult feelings, to both honor and resist them. Doctors have a supply of emotions and insights about body and mind, and to offer access to this storehouse is the gift, the art of medicine, just as memories are the gifts we receive from loved ones.

Betrayed, in robe and slippers, the patient believes that the more he tells the less you know. Caregivers can't bear *not* knowing the secrets of illness. As they accept that illness is a crime and a trial, caregivers represent the right to ordinary decency. They can admit to losses and keep the loved one in sight even as he believes he is vanishing. Away from the fierce and tender confidence of doctors, what caregivers can do is feel afraid for the patient.

What patients can do during illness, within its ever-shrinking perimeter, is speak of any or all of the four feelings I've presented here. The voice is a way of extending outward, of occupying space beyond the body. Whether confessing, begging, commanding, blessing, consenting, or cursing, the patient can describe the secret aspects of illness that are specific to him. These are the modes of survival. As Charlie pointed out to me, the spray paint on the underpass nearest my hospital reads: SILENCE = DEATH.

Though Richard was dying, things inside him were alive and unrelenting. When he was well, Richard exuded confidence and self-satisfaction, and strangely, in bed, wasted and dry-lipped, he still came across as a man with things to do, a man whose work might be temporarily suspended but who I expected at any moment would be ready to converse and argue about the work that others should be doing until he fully got back to his own. Instead, whispering, he baited me to tell him how he looked. Emaciated, he looked elegant. His features had sharpened as he lost weight. His hair, combed back, lay flat as piano wires. His skin had thinned, so that his veins looked like embroidery. As an artist, he was accustomed to his own company, but he considered me an ever-ready audience. He had always been a talker, rigorous and wandering, and when he wasn't at my throat, I loved his company. I rarely felt a greater joy than when I said something that interested him.

I provided the optimism, the myth of Richard's final illness. Even when he was bed-bound and groaning near the end, I simply imagined moving his bed from place to place over the years to come so that he could be with us; even if he didn't participate, he could be present. All along, he'd wanted his doctors to be wrong and me, his private physician, to be right in speculating that the cancer was a slow-growing type; maybe it was so slow it would halt. I'd gone along with this notion because it was a pleasant thought. I forgot everything I knew and just cared for him. His doctors had attended to all the tedious particulars of his illness, and now the only thing on my mind was missing him. It was good to see him even in his bottoming state. Richard had been so close to my life that I never bothered to imagine he'd be gone, that I would have to go through his closet of clothes with my sister, deciding which to keep and which to give away. As much as illness is a state of hyper-reality, it also contains illusion.

Sometimes it is disbelief that keeps a patient alive. He believes that death has nothing to do with him. On a bridge suspended over the void, he doesn't worry about the bridge collapsing. Even when most of one's ordinary expectations have collapsed, there is no existential disaster at hand. One's sense of place has tumbled, one's sense of time has been interfered with, but the body has the capacity to make infinite, tiny, incomprehensible adjustments, and the patient has every right to believe it will continue to do so. Illness involves a constant set of corrections—move this way, breathe that way—in order to survive.

The room was small, the floors dark wood, the built-in bookcases shadowed. Death always resigns the survivors to the littleness of life. If Richard was imagining death, he never spoke of it. So he looked past it. He wanted to be properly taken care of after he was gone: would his ashes fit in the Chinese funerary urn? I was silent—the silence of being afraid to look. It took considerable effort for Richard to speak at all, and he would nod off when conversation became difficult.

I found myself looking at the clock in the kitchen often. It held only remaining time now; the minutes were limited, and I wanted to count every one. The moment Richard drifted off to sleep, I wiped the corners of his mouth and took away the straws he had dropped from the water cups that lay along his chest slewed in sheets. I remembered a character in a story I once read saying, "Once you start napping, you're dead in no time."

Amid the sadness, we all tried to keep good humor between the savage hours when he choked, slid off the bed, got his catheter tangled in the blankets, and grew more and more disoriented. I pretended to be brave to mirror his actual braveness. I held his cool hands and thought of the treasures he would leave behind, and the anguish.

Illness summons the memory of any prior illnesses we might have had or observed, the illnesses of our friends and relatives. What we visualize when we are sick is the man in green arriving with his stretcher to take Mother to her hysterectomy; our younger brother in bed having the osteomyelitis of his ankle debrided by the visiting surgeon; our uncle's stamina and gathering weakness during the immense exhaustion of chemotherapy; Grandmother's hips starved to points by tumors as she covers her eyes with her hand, a gesture of grief and frustration and weariness; our aunt, confused and swollen, lifting her purple foot and asking, "Does this seem right to you?" These important sights and ominous chords are burned into our minds.

Being in the presence of illness has always moved me. Seeing Richard, I thought again of my father dying in an emergency room I was not allowed in—and then he was gone. I was a thirteen-year-old boy; the waiting room was white; until that night I had believed that nothing would ever change. Illness reawakens in me memories of these hard times. Near illness, I always have the sense that time is short, that things I depend on disappear. Every patient reattaches me to those moments when I got a close-up look at Richard's resistance. Illness slows me down, and I become self-consciously searching.

The day before Richard died, the cats Cecil and Alice, who had always cuddled with him, who loved him and had never betrayed him, jumped off the bed and wouldn't return. Even when I tried to replace them in the crook of his arm, they wouldn't stay. They knew.

My dread, his end. Travel sometimes provides a permanence. For some patients, the trip will go on forever. They die.

I have considered the worth of Richard's life and steadily tried to become the artist he was, and the doctor he would have wanted. "Pay attention, this is our endless and proper work," the poet Mary Oliver has written. I try. Illness is never simply a technical problem to be solved. It is personal business. It involves unease and decent concern, jargon and gesture, but never should it include turning aside from need. Richard's illness was a reeducation for me as a doctor and friend. But for the patient, illness is as unshareable as a book read alone on the last train out of town.

Acknowledgments

To manage two careers, writing and doctoring, means surrounding oneself with many generous and supportive friends. My colleagues at Brown Medical School, particularly those in the Division of General Internal Medicine, have shared their love of medicine, their energy in serving the vulnerable, and their many clinical insights. Over the years my patients at Rhode Island Hospital have told me everything, put up with my nosiness, and taught me how to accept the pleasure that comes from helping. Thanks to Martin and Judy Shepard who, for more than a decade, have nurtured my career as a novelist and encouraged me to write the stories I care about. If Betsy Lerner were merely brilliantly inventive, it would be enough. But she is kind, patient, and available, all a writer needs in a teacher. My superior editor, Henry Ferris, made me believe that this project was important, while at the same time making every page better with his sharp mind and blue pen. Tobias and Alexander, my wonderful sons, make me laugh with their lists of silly diagnoses and their bloodthirsty Boggle skills; I would rather eat dinner with them than with anyone else. Hester, my favorite author, has taught me kindness, what it means to have an open heart, and all there is to know about love, the attributes one needs for doctoring, writing, and living a life.